from GENETICS to GENE THERAPY

UCL Molecular Pathology series

Editor: **D.S. Latchman**
Department of Molecular Pathology, University College London, The Windeyer Building, 46 Cleveland Street, London W1P 6DB, UK

From Genetics to Gene Therapy

Autoimmune Diseases: focus on Sjögren's syndrome

RJB2/90

from GENETICS to GENE THERAPY

the molecular pathology of human disease

David S. Latchman
Department of Molecular Pathology, University College London, London, UK

BIOS
SCIENTIFIC
PUBLISHERS

First published 1994

A CIP catalogue record for this book is available from the British Library.

ISBN 1 872748 36 8

BIOS Scientific Publishers Ltd
St Thomas House, Becket Street, Oxford OX1 1SJ, UK
Tel. +44 (0)865 726286. Fax +44 (0)865 246823

DISTRIBUTORS

Australia and New Zealand
DA Information Services
648 Whitehorse Road, Mitcham
Victoria 3132

India
Viva Books Private Limited
4346/4C Ansari Road
New Delhi 110002

Singapore and South East Asia
Toppan Company (S) PTE Ltd
38 Liu Fang Road, Jurong
Singapore 2262

USA and Canada
Books International Inc
PO Box 605, Herndon,
VA 22070

Typeset by Ann Buchan (Typesetters), Shepperton, UK.
Printed by Information Press Ltd, Oxford, UK.

Contents

Contributors

Bishop, A.E. Department of Histochemistry, Royal Postgraduate Medical School, Hammersmith Hospital, Du Cane Road, London W12 0NN, UK

Brown, S.C. Department of Experimental Pathology, UMDS Guys Hospital, London Bridge, London SE1 9RT, UK

Collins, M.K.L. Chester Beatty Laboratories, Leukaemia Research Fund Centre, Institute of Cancer Research, 237 Fulham Road, London SW3 6JB, UK

Darley, D.L. Department of Haematology, University College of Wales College of Medicine, Cardiff CF4 4XW, UK

Dickson, G. Department of Experimental Pathology, UMDS Guys Hospital, London Bridge, London SE1 9RT, UK

Greaves, M.F. Chester Beatty Laboratories, Leukaemia Research Fund Centre, Institute of Cancer Research, 237 Fulham Road, London SW3 6JB, UK

Kinnon, C. Molecular Immunology Unit, Division of Cell and Molecular Biology, Institute of Child Health, 30 Guilford Street, London WC1N 1EH, UK

Landon, M. Department of Biochemistry, University of Nottingham Medical School, Queens Medical Centre, Nottingham NG7 2UH, UK

Latchman, D.S. Department of Molecular Pathology, University College London, The Windeyer Building, 46 Cleveland Street, London W1P 6DB, UK

Lowe, J. Department of Pathology, University of Nottingham Medical School, Queens Medical Centre, Nottingham NG7 2UH, UK

Lowenstein, P.R. Department of Physiology, University College of Wales College of Medicine, Cardiff CF4 4XW, UK

Mayer, R.J. Department of Biochemistry, University of Nottingham

Medical School, Queens Medical Centre, Nottingham NG7 2UH, UK

Nabel, E.G. Howard Hughes Medical Institute Research Laboratories, The University of Michigan Medical Center, MSRB I, 1150 W. Medical Center Drive, Ann Arbor, MI 48109-0650, USA

Nabel. G.J. Howard Hughes Medical Institute Research Laboratories, The University of Michigan Medical Center, MSRB I, Room 4510, 1150 W. Medical Center Drive, Ann Arbor, MI 48109-0650, USA

Polak, J.M. Department of Histochemistry, Royal Postgraduate Medical School, Hammersmith Hospital, Du Cane Road, London W12 0NN, UK

Scott, J. MRC Molecular Medicine Group, Royal Postgraduate Medical School, Hammersmith Hospital, Du Cane Road, London W12 0NN, UK

Stratton, M.R. Haddow Laboratories, Institute of Cancer Research, Sutton, Surrey SM2 5NG, UK

Tipler, C. Department of Biochemistry, University of Nottingham Medical School, Queens Medical Centre, Nottingham NG7 2UH, UK

Weiss, R.A. Chester Beatty Laboratories, Institute of Cancer Research, 237 Fulham Road, London SW3 6JB, UK

Wilkinson, G.W.G. Department of Medicine, University College of Wales College of Medicine, Cardiff CF4 4XW, UK

Abbreviations

AAV	adeno-associated virus
ABC	avidin–biotin–peroxidase
Ad	adenovirus
AD	Alzheimer's disease
ADA	adenosine deaminase
AIDS	acquired immunodeficiency syndrome
ALL	acute lymphoblastic leukaemia
ALV	avian leukosis virus
apo	apolipoprotein
APP	amyloid precursor protein
ARD	acute respiratory disease
AT	ataxia-telangiectasia
BHK	baby hamster kidney
BLV	bovine leukosis virus
BMD	Becker muscular dystrophy
BrdU	bromodeoxyuridine
BSE	bovine spongiform encephalopathy
CD40L	CD40 ligand
CGD	X-linked chronic granulomatous disease
CGRP	calcitonin gene-related peptide
CI	confidence interval
CJD	Creutzfeld–Jakob disease
CML	chronic myeloid leukaemia
CMV	cytomegalovirus
CNS	central nervous system
DAGC	dystrophin-associated glycoprotein complex
DMD	Duchenne muscular dystrophy
DRP	dystrophin-related protein
EC	enterochromaffin
ER	endoplasmic reticulum
FAD	familial Alzheimer's disease
FeLV	feline leukaemia virus
FGF	fibroblast growth factor
FISH	fluorescence *in situ* hybridization

GALV	gibbon ape leukaemia virus
GSS	Gerstmann–Straussler–Scheinker disease
α-HCG	α human chorionic gonadotrophin
HDL	high density lipoprotein
HIGMX1	hyper IgM syndrome
HIV	human immunodeficiency virus
hsp	heat-shock protein
HSV	herpes simplex virus
HSV-tk	HSV thymidine kinase
HTLV	Human T-cell leukaemia virus
IDL	intermediate density lipoprotein
IE	immediate–early
IL2RG	interleukin-2 receptor gene
K^s	kanomycin-sensitive
LAT	latency-associated transcripts
LBD	Lewy body dementia
L-APP	leukocyte-derived amyloid precursor protein
(V)LDL	(very) low density lipoprotein
LDLR	low density lipoprotein receptor
LMP	latent membrane protein
LTR	long terminal repeat
MAP	microtubule-associated protein
MEN	multiple endocrine neoplasia
MHC	major histocompatability complex
Mo MLV	Moloney murine leukaemia virus
MTC	medullary thyroid carcinoma
NBT	nitroblue tetrazolium
NGF	nerve growth factor
ORF	open reading frame
PCNA	proliferating cell nuclear antigen
PCR	polymerase chain reaction
PD	Parkinson's disease
PDGF	platelet-derived growth factor
PDI	protein disulphide isomerase
PFU	plaque-forming units
β-PPT	β-pre-protachykinin
PrP	prion protein
RER	rough endoplasmic reticulum
RFLP	restriction fragment length polymorphism
RR	relative risk
RSV	Rous sarcoma virus

SCID(X1)	(X-linked) severe combined immune deficiency
SH	*src* homology
SIV	simian immunodeficiency virus
SSCP	single-stranded conformational polymorphism
SU protein	surface protein
SV40	simian virus 40
Tag	transforming antigen
TILS	tumour-infiltrating lymphocytes
TM	transmembrane
TNF	tumour necrosis factor
UTR	untranslated region
VIP	vasoactive intestinal peptide
VSV	vesicular stomatitis virus
WAS	Wiskott–Aldrich syndrome
WHO	World Health Organization
XLP	X-linked lymphoproliferative syndrome
YAC	yeast artificial chromosome

Preface

As discussed in Chapter 1, the increasing application of molecular biology to medical problems has led to the creation of a new subject, molecular pathology, which uses molecular techniques to advance the understanding, diagnosis and ultimate therapy of human disease. Recognizing the importance of this subject, UCL Medical School has created a new Department of Molecular Pathology by bringing together the existing Department of Chemical Pathology and the Medical Molecular Biology Unit (formerly part of the Department of Biochemistry and Molecular Biology).

To celebrate the formation of the new Department a 1-day meeting entitled 'From Genetics to Gene Therapy' was held on 14 December 1992, at which a large audience was inspired by talks from a variety of distinguished speakers. This volume is based on the meeting and contains contributions provided by the invited speakers. However, the very limited time available at the meeting necessitated the omission of some important topics in this area and, for this reason, the contributions by the invited speakers are now complemented by a series of additional chapters which have been especially commissioned for this volume. I have also included a chapter (Chapter 13) based on my inaugural lecture as Professor of Molecular Pathology at UCL, which was delivered on the day after the meeting.

The diverse range of topics covered in the book includes infectious diseases, simple genetic diseases and multifactorial diseases, providing an overview of the field, with particular emphasis on the move towards gene therapy which is now occurring in a number of diseases. Thus, the contribution to the meeting by James Scott on apolipoprotein B and coronary heart disease is complemented by additional chapters on the genetic defects in X-linked immunodeficiency diseases and in Duchenne muscular dystrophy, which focus, in particular, on the gene therapy approach to these diseases. Similarly, the important topic of cancer is dealt with in several chapters which cover topics ranging from cancer genetics to cancer gene therapy. A major focus at the meeting was the technology of gene therapy, and the contributions in this area by Robin Weiss, on retroviral receptors, and Gary Nabel, on direct gene transfer, are supplemented by an overview of the viral vectors available for this purpose. The ultimate target of gene therapy will be the neurodegenerative diseases, and an additional chapter is now included on this topic in order to complement my inaugural lecture on herpes simplex virus which, as well as being an important disease-causing agent, may ultimately be of use as a vector for gene therapy in these diseases. I would like to express my gratitude to all the authors for their excellent contributions to this volume.

The initial symposium and this volume are intended to serve as an introduction to the field. In subsequent years there will be further annual symposia focussing on the molecular pathology of a single disease or group of diseases. The volumes based on these symposia will therefore provide a detailed picture of the impact of molecular biology in a particular area. The second annual symposium dealt with the 'Molecular Pathology of Autoimmunity: Focus on Sjögren's Syndrome', and was organized by Professor David Isenberg and Dr Angela Horsfall, whilst the third symposium will be organized by Dr Sean Thomas and will deal with the molecular pathology of malignant disease, with special reference to cell cycle control.

The success of this first volume and the continuing series of volumes is greatly dependent on the efficiency of the staff at BIOS Scientific Publishers, and I would like to express my thanks to them, as well as to all the contributors, for helping to launch the new series in such an appropriate manner.

David Latchman

Foreword

Over recent years there has been an enormous expansion in the application of the technologies of molecular and cell biology to the study of human disease. This new and exciting field is moving so rapidly that it is extremely difficult even for those who are working in it to keep up with any areas other than their own particular research interests, and the problem of keeping abreast with its major developments is even more difficult for busy academic clinicians. But although this reductionist approach to human pathology may well, in the longer term, have to be replaced by a more integrative approach, in which the actions of individual genes are seen in the context of intact cells and individuals, it is likely that the molecular sciences will dominate much of medical research for the immediate future.

This book contains a series of general reviews on different aspects of molecular pathology, all written by workers who are active in the field. The different topics are presented in a way which should make them immediately available to non-specialists and to those in the different branches of the clinical and basic biomedical sciences, who wish to keep abreast of these fast-moving developments. Given the state of the art, the book concentrates on molecular pathology but it also gives some inkling of the thinking which is behind efforts to transfer human genes for therapeutic purposes. While the immediate prospects of successful gene therapy may have been rather overstated, there now seems little doubt that this general approach will offer a wide spectrum of therapeutic options across many branches of medical practice in the future. Thus, this book should be read by young medical or non-medical scientists who are thinking about a career in this field, or those who wish to find out what is going on and about the current state of progress in some of the more important areas of molecular medicine.

D.J. Weatherall

1

What is molecular pathology?

D.S. Latchman
Department of Molecular Pathology, University College London Medical School, London, UK

Pathology has a long and honourable history as the science of disease. Over the years, the subject became divided into a number of sub-disciplines. The medical microbiologist studied the nature of infectious disease-causing organisms, the histopathologist studied the structure of diseased tissue, whilst the chemical pathologist studied the changes in their chemical composition.

In the last 20 years, however, pathology has been transformed by the impact of the molecular biology revolution, which has permeated all these different aspects of pathology. This has led to the development of a new subject, molecular pathology, which has as its aim the use of the techniques of molecular biology to improve the understanding, diagnosis and therapy of human disease.

The impact of the new technology has been particularly marked in the medical microbiology area. Initial advances involving the detection of, for example, viral nucleic acids by Southern blotting or *in situ* hybridization (Teo, 1990), have now been supplemented by the polymerase chain reaction (PCR; Saiki *et al.*, 1985), whose extreme sensitivity is rendering it of immense value in the diagnostic laboratory (Kwok *et al.*, 1988). The recent development of the PCR–*in situ* hybridization technique, which combines the cellular localization provided by the *in situ* technique with the sensitivity of PCR, is likely to lead to further advances in this area (Nuovo *et al.*, 1991). Indeed, this technique has already led to an enhanced understanding of the pathogenesis of HIV infection, allowing the detection of a very large pool of latently infected lymphocytes and macrophages, even at early stages of infection (Embretson *et al.*, 1993).

This impact of molecular biology on both our understanding of disease and its diagnosis is also seen in the area of human inherited diseases in which the defective protein had not been readily identifiable in studies using conventional biochemical techniques. In these cases, prenatal diagnosis of an at risk fetus was greatly facilitated by the use of restriction fragment length polymorphisms (RFLPs), which provided a whole new array of genetic markers that could be used to follow the inheritance of the uncharacterized disease locus to which they were linked. Subsequently, minisatellites (Jeffreys *et al.*, 1985) or microsatellites combined with PCR (Litt, 1991) provided a further array of polymorphic markers for use in this way.

An early example of the use of this technique, involving the neurological disease Huntington's chorea (Gusella *et al.*, 1983) resulted some 10 years later in the identification of the defective gene itself (Huntington's Disease Collaborative Research Group, 1993). Thus, once a linked marker has been identified, it is possible to isolate further markers which are more closely linked to the disease gene and, ultimately, to identify the disease gene itself. Although such positional cloning is a laborious and time-consuming process, it does not normally take as long as in the case of Huntington's chorea, and a number of disease loci have been identified by this so-called 'reverse genetics' approach to supplement its early successes involving the genes responsible for cystic fibrosis (Riordan *et al.*, 1989) and Duchenne muscular dystrophy (Koenig *et al.*, 1987).

Of course, the identification of the defective gene greatly increases the efficiency of prenatal diagnosis, which can be carried out with probes for the gene itself or for very closely linked markers rather than with markers which may be subject to recombination, leading to a potentially tragic incorrect diagnosis. More importantly, however, the isolation of the gene itself allows studies of the molecular pathology of the disease to move forward from its diagnosis to its understanding. Thus, once the gene is available, the nature of the mutation(s) which result in the disease can be studied in exactly the same way as was done for the genes encoding defective proteins, identified by biochemical means in specific diseases, as in the case of the various mutations in the β-globin genes which occur in the β-thalassaemias (Weatherall, 1983).

As well as identifying the expected gene deletions or alterations in encoded amino acids which affect the protein itself, these studies have revealed a plethora of other mutations which affect the expression of the gene in question, resulting in a failure of protein production. Thus, for example, in haemophilia B, a mutation in the factor IX gene promoter abolishes the binding of the stimulatory transcription factor, C/EBP, leading to a failure of gene transcription and, consequently, of protein production (Crossley and Brownlee, 1990). Similarly, some cases of β-thalassaemia are caused by mutations which affect the splicing of the primary RNA transcript, resulting in the production of aberrant RNA species which are rapidly degraded and which, in any case, could not encode a functional β-globin protein (Treisman *et al.*, 1983).

In addition to structural and regulatory mutations, the positional cloning of disease genes has revealed a novel type of mutation. Thus, several genes, including those defective in diseases such as fragile X syndrome, myotonic dystrophy and Huntington's chorea, contain multiple copies of a 3-bp sequence which is different in each case (for review, see Richards and Sutherland, 1992). In normal individuals, there are only a small number of contiguous copies of these triplet repeats, whereas in individuals suffering from the disease the repeated sequence becomes highly amplified. As such amplification occurs progressively in several generations, these so-called dynamic mutations neatly explain the genetic phenomenon of anticipation, whereby the severity of the disease increases in successive generations of affected pedigrees. It remains unclear, however, why such repeat amplifications cause the disease, particularly as they are often not located in the protein-coding region of the gene. Similarly, it is unclear why the defective copy of the gene can often prevent the function of the normal allele, resulting in the dominant nature of these mutations which is observed in, for example, Huntington's chorea. The understanding of this effect is likely to throw considerable light on the cause of these diseases and on the role of the normal gene product in cellular function.

Ultimately, however, the aim of molecular pathology must be the improved therapy of human disease rather than simply its better understanding or better diagnosis. In this area, initial studies focussed on the expression of human genes in bacteria or eukaryotic cells, leading to the large-scale production of human proteins such as insulin or growth hormone (see, for example, Goeddel *et al.*, 1979). Such recombinant proteins can then be used in the therapy of diseases where these substances are not produced. This is of particular value with an example such as growth hormone, where animal sources are not effective in humans and the protein had to be purified from human cadavers, with consequent risk of infection. To date, some 17 recombinant proteins have been licensed by the FDA for use in humans in this way.

As well as being of use in the treatment of genetic diseases, recombinant proteins can be used in the production of effective vaccines against infectious diseases. Thus, for example, the expression of the hepatitis B virus major surface antigen in yeast provided a large source of the protein which is now widely used as the immunogen to provide an effective vaccine against the severe debilitating disease caused by infection with this virus (Gerety, 1988).

Although of particular value in the development of vaccines against infectious diseases, the use of recombinant proteins to treat genetic diseases suffer from the defect that repeated injections are necessary to replace the protein as it is naturally turned over. The obvious solution to this problem is to insert an intact copy of the gene encoding the protein into the cells which contain only a defective gene. The normal gene would then produce a continuous supply of the appropriate protein, obviating the need for a continuous supply of exogenous protein.

For obvious reasons, this gene therapy approach has attracted considerable attention. A number of different gene delivery systems, including various types of viral vectors and physical transfer methods, are now being developed for this purpose, and clinical trials of the efficacy of this procedure have already begun for several disorders (for review, see Miller, 1992). Such an approach clearly has much to offer, not only in genetic diseases, but also as a means of introducing target antigens into cancer cells to stimulate immune rejection or to express dominant inhibitory proteins to block the progress of viral infections.

It should be noted, however, that the route which leads from genetic mapping of a defect to isolation of a defective gene and on to therapy with the intact gene is fraught with danger if carried out in the absence of information as to the role of the gene in normal cellular function. Thus, the recent cloning of the defective gene in X-linked agammaglobulinaemia reveals it to be a member of the *src* family of proto-oncogenes, which encode growth-regulating tyrosine kinases (Vetrie *et al.*, 1993; see Chapter 3). This suggests that care should be taken in the use of this gene in gene therapy procedures. Similarly, gene therapy in Huntington's disease will be impossible until it is understood why the presence of defective protein causes the disease in heterozygotes with an intact functional copy of the appropriate gene.

These considerations indicate that an understanding of the normal role of specific gene products, and the manner in which their absence or alteration results in disease, is likely to be the most important contribution which molecular pathology can make to the treatment of human disease, allowing the design of novel therapies both of the genetic and pharmaceutical type. In this manner, molecular pathology will fulfil its aim of allowing improved diagnosis and understanding of specific human diseases to result ultimately in their improved therapy.

References

Crossley M and Brownlee GG. (1990) Disruption of a C/EBP binding site in the factor IX promoter is associated with haemophilia B. *Nature*, **345**, 444–446.

Embretson J, Zupancic M, Ribas JL, Burke A, Racz P, Tenner-Racz K and Haase AT. (1993) Massive covert infection of helper T lymphocytes and macrophages by HIV during the incubation period of AIDS. *Nature*, **362**, 359–362.

Gerety RJ. (1988) Recombinant hepatitis B vaccines. In: *Viral Hepatitis and Liver Disease* (ed. AJ Zukerman). Alan R. Liss, New York, pp. 1017–1024.

Goeddel DV, Heyneker HL, Hozumi T *et al.* (1979) Direct expression in *Escherichia coli* of a DNA sequence coding for human growth hormone. *Nature*, **281**, 544–548.

Gusella JF, Wexler NS, Conneally PM *et al.* (1983) A polymorphic DNA marker linked to Huntington's disease. *Nature*, **306**, 234–238.

Huntington's Disease Collaborative Research Group (1993) A novel gene containing a trinucleotide repeat that is expanded and unstable on Huntington's disease chromosomes. *Cell*, **72**, 971–983.

Jeffreys AJ, Wilson W and Thein SL. (1985) Hypervariable mini satellite regions in human DNA. *Nature*, **314**, 67–73.

Koenig M, Hoffman EP, Bertelson CJ, Monaco AP, Feener C and Kunkel LM. (1987) Complete cloning of the Duchenne muscular dystrophy (DMD) cDNA and preliminary genomic organization of the DMD gene in normal and affected individuals. *Cell*, **50**, 509–517.

Kwok S, Ehrlick G, Poiesz B, Kalish R and Sninsky J. (1988) Enzymatic amplification of HTLV-1 viral sequences from peripheral blood mononuclear cells and infected tissues. *Blood*, **72**, 1117–1123.

Litt M. (1991) PCR of TG microsatellites. In: *PCR: a Practical Approach* (eds MJ McPherson, P Quirke and GR Taylor). IRL Press, Oxford, pp. 85–99.

Miller AD. (1992) Human gene therapy comes of age. *Nature*, **357**, 455–460.

Nuovo G, MacConnell P, Forde A and Delvenne P. (1991) Detection of human papillomavirus DNA in formalin fixed tissues by *in situ* hybridization after amplification by polymerase chain reaction. *Am. J. Pathol.* **139**, 847–850.

Richards RI and Sutherland GR. (1992) Dynamic mutations: a new class of mutations causing human disease. *Cell*, **70**, 709–712.

Riordan JR, Rommens JM, Kerem B-S *et al.* (1989) Identification of the cystic fibrosis gene: cloning and characterization of complementary DNA. *Science*, **245**, 1066–1073.

Saiki RK, Scharf S, Faloona F, Mullis KB, Horn GT, Erlich HA and Arnheim N. (1985) Enzymatic amplification of β-globin genomic sequences and restriction site analysis for diagnosis of sickle cell anemia. *Science*, **230**, 1350–1354.

Teo CG. (1990) *In situ* hybridization in virology. In: In Situ *Hybridization Principles and Practice* (eds JM Polak and JOD McGee). Oxford University Press, Oxford, pp. 125–147.

Treisman R, Orkin SH and Maniatis T. (1983) Specific transcription and RNA splicing defects in five cloned β-thalassaemia genes. *Nature*, **302**, 591–596.

Vetrie D, Vorechovsky I, Sideras P *et al.* (1993) The gene involved in X-linked agammaglobulinaemia is a member of the *src* family of protein-tyrosine kinases. *Nature*, **361**, 226–233.

Weatherall DJ. (ed.) (1983) *The Thalassaemias*. Churchill Livingstone, Edinburgh.

2

Apolipoprotein B and coronary heart disease

J. Scott
Department of Medicine, Royal Postgraduate Medical School, Hammersmith Hospital, London, UK

Summary

Structural domains in apolipoprotein (apo) B100 that mediate lipoprotein assembly, secretion and clearance have been delineated. The regulation of apo B-containing lipoprotein biosynthesis is entirely post-translational. Lipid availability channels nascent apo B100 between pre-secretory degradation and secretion as a lipoprotein particle.

Apo B48 has been shown to be co-linear with the amino-terminal half of apo B100 and to be produced by a novel mRNA editing process, which inserts a stop translation codon. By development of an *in vitro* editing system, the mechanism of, and sequence requirements for, this process have been delineated. The apo B-mRNA-editing enzyme is a site-specific cytidine deaminase that recognizes an RNA sequence at a fixed distance downstream of the edited site. The RNA recognition component of the editing enzyme has been identified as a 60 kDa protein with a well delineated RNA recognition motif.

Apo B is the major cholesterol and triglyceride transport protein in the blood. As such, apo B has a major role in determining the risk of cardiovascular disease. This chapter is concerned with the structure and function of apo B in relation to metabolic and genetic factors that affect the circulating levels of apo B-containing lipoproteins.

1. The metabolism of apo B-containing lipoprotein

Circulating lipoproteins consist of an oily droplet with a hydrophobic core of neutral lipid (triglyceride and cholesteryl-ester). This is surrounded by a monolayer of polar phospholipid and free cholesterol which is stabilized by the amphipathic binding of apolipoproteins to the surface of the particle. The composition of lipoprotein particles is documented elsewhere (Scott, 1989a,b, 1990, and literature cited therein).

Two pathways of apo B-containing lipoprotein metabolism have been described. The exogenous pathway is responsible for the transport of dietary lipid and uses apo B48. Apo B48 is the intestinal form of apo B in humans and is designated on the centile system because it is 48% of the size of hepatic apo B100 on SDS–PAGE. The endogenous pathway is necessary for the transport of triglyceride and cholesterol synthesized in the liver and involves apo B100 in humans, and both apo B100 and apo B48 in rodents.

1.1. Apo B48 and dietary lipid absorption (the exogenous pathway)

Within the intestinal absorptive cell, dietary lipid and fat-soluble vitamins are packaged into large (100–500 nm) triglyceride-rich lipoprotein particles called chylomicrons. The major apolipoprotein of chylomicrons is apo B48, with smaller amounts of the apo C peptides (apo CI, apo CII and apo CIII) and of apo AIV. Chylomicrons are secreted into intestinal lymph and enter the blood through the thoracic duct for transport to peripheral capillaries, where core triglycerides are hydrolysed by endothelial cell lipoprotein lipase and released fatty acids are taken up by fat and skeletal muscle cells. The loss of triglycerides released from the chylomicron produces a remnant particle, relatively enriched in cholesteryl-esters and containing apo B48 and apo E. This remnant is transported to the liver where it is taken up by both the low density lipoprotein (LDL) receptor and a specific chylomicron remnant receptor.

1.2. Apo B100 and endogenous lipid transport (the endogenous pathway)

The liver synthesizes triglyceride and cholesterol which, together with residual dietary lipid and fat-soluble vitamins, are incorporated into very low density lipoprotein (VLDL) and secreted into the circulation. VLDL contains apo B100 as its major lipoprotein, together with apo C and apo E. VLDL is transported to the periphery where the triglyceride is hydrolysed by the same lipoprotein lipase that catabolizes chylomicrons. After removal of core triglyceride, VLDL remnants, called intermediate density lipoprotein (IDL), undergo two distinct fates. Some of the IDL is cleared directly by the liver through interaction between apo E and the LDL receptor. The remainder is acted on by hepatic lipase in the hepatic sinusoids. All of the residual triglyceride is hydrolysed and replaced with cholesteryl-ester. In the course of this conversion all the apolipoproteins, except for apo B100, are lost to other lipoprotein particles. The result is a transformation of VLDL remnant particles into cholesteryl-ester-rich LDL, where apo B is the sole lipoprotein. LDL carries 60–70% of plasma cholesterol and functions in the delivery of cholesterol to peripheral tissues, where it is required for membrane and steroid hormone biosynthesis, and to the liver for further metabolism or excretion in bile. Apo B100 is the ligand that mediates the cellular uptake of LDL by the LDL receptor pathway.

2. The structural and functional domains of apo B100

Apo B100-containing lipoproteins have a single apo B molecule each. The functions of apo B100 are as a nucleus for the assembly of apo B-containing lipoproteins in the liver, to stabilize apo B-containing lipoprotein during the metabolic transitions that take place in the circulation and as a ligand that mediates the clearance of LDL from the circulation by the LDL receptor pathway.

2.1. The LDL receptor-binding domain of apo B100

Using a panel of monoclonal antibodies raised against apo B in the context of LDL and delipidated apo B, the domain that binds to the LDL receptor has been localized to the carboxy-terminal half of apo B100 (*Figure 1*). The localization of this domain was confirmed with the identification of an arginine (CGG) to glutamine (CAG) mutation at position 3500 that causes moderately severe 'familial' hypercholesterolaemia (familial defective apo B) in which apo B binds to the LDL receptor (*Figure 1*; Innerarity, 1990).

2.2. Lipid-binding structures in apo B, lipoprotein assembly and plasma lipoprotein integrity

Predictions of the hydrophobicity and secondary structure of apo B100, in conjunction with proteolysis and lipid-binding studies, indicate that apo B100 is capable of binding lipid throughout its very considerable

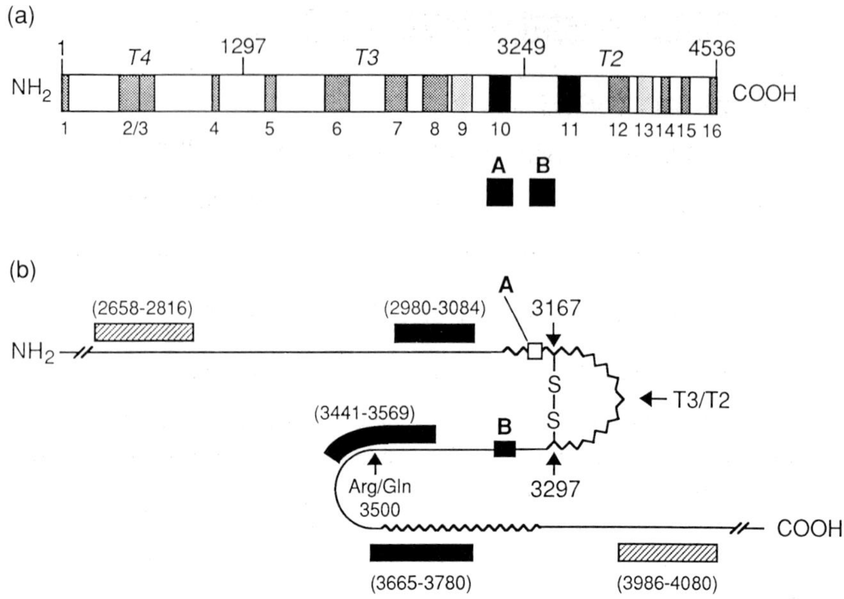

Figure 1. Localization (a) and organization (b) of the LDL receptor-binding domain of apo B100. A panel of monoclonal antibodies have been mapped to apo B100. Antibodies that block the binding of apo B to the LDL receptor have epitopes that cluster around two basic peptides in the carboxy-terminal half of apo B100. Solid boxes are complete blocking antibodies; stippled boxes are partial blocking antibodies.

length (Milne *et al.*, 1989; Pease *et al.*, 1990). The sequence of the apo B molecule alternates between hydrophobic and hydrophilic structure. Throughout the sequence are short peptides predicted to form either amphipathic α-helices or amphipathic β-strands. In addition, two domains containing extended amphipathic α-helices and four others predicted to form amphipathic β-sheets are found in apo B (*Figure 2*). An additional structure predicted to exist in apo B is the covalent association of the protein with fatty acid; this would also be predicted to bind lipid. The lipid-binding structures in apo B have in common the feature that none except for the signal peptide would be predicted to cross a cellular membrane. Thus, apo B is a huge amphipathic molecule predicted to be associated with the inner leaflet of the endoplasmic reticulum (ER) during lipoprotein assembly and with the monolayer of phospholipid and free cholesterol at the surface of apo B-containing lipoprotein particles.

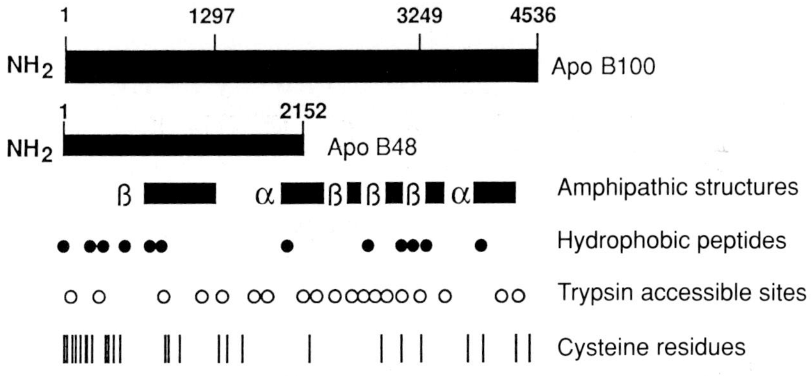

Figure 2. Disposition of lipid-binding structures in apo B100.

The study of genetic disorders that affect the assembly and secretion of apo B-containing lipoproteins has been important for understanding the role of apo B and of proteins determined by other genetic loci in this process. The Mendelian co-dominant disorder hypobetalipoproteinaemia has been shown to be a defect of the apo B gene in which truncated variants of apo B, corresponding to apo B31 on the centile system, are capable of forming buoyant lipoprotein particles, whereas forms of apo B, apo B29 or apo B25 have not been detected in the circulation (Innerarity, 1990; Scott, 1989a,b). These observations have led to the view that there is a certain minimum size corresponding approximately to the amino-terminal one-third of apo B100 that is necessary for the formation and secretion of buoyant lipoproteins. This view has been confirmed by the expression of truncated forms of apo B in human and rodent hepatoma cell lines (Graham *et al.*, 1991; White *et al.*, 1992; Yao *et al.*, 1991). Results *in vitro* are consistent with the conclusion that truncated forms of apo B below the size of apo B48 are not secreted as VLDL size particles but rather as particles in the LDL and HDL density range, depending on their size. Below the size of apo B23, truncated forms of apo B are secreted from hepatoma cells but do not form lipoprotein particles and float at a density of soluble protein.

In summary, apo B100 can be divided into two basic functional domains: the carboxy-terminal which, although lipid binding in structure, is also the region that interacts with the LDL receptor, and the amino-terminal half, which contains all the structural information necessary for the assembly and secretion of lipoprotein particles.

3. Apo B-containing lipoprotein assembly and control of apo B secretion

3.1. The assembly of apo B-containing lipoproteins

Studies of ultrastructure, together with pulse-chase and cell fractionation studies and, more recently, studies involving *in vitro* translation and translocation of truncated forms of apo B, have provided us with a picture of the kinetics and molecular mechanisms involved in the assembly of apo B-containing lipoproteins (*Figure 3*).

Apo B is so large a protein that it is predicted to take some 15 minutes to translate. Apo B undergoes a variety of co- or post-translational modifications, including intrachain and interchain disulphide bonding with apo(a), N-linked glycosylation, phosphorylation of serine, threonine and tyrosine, and fatty acylation. Immediately after translation is complete, nascent lipoprotein particles can be detected in the lumen of the ER (Borén *et al.*, 1992). At this stage, apo B has received neutral lipid, phospholipid and free cholesterol from their sites of synthesis in the outer leaflet of the ER. Although a matter of controversy, the available structural predictions on apo B and *in vitro*

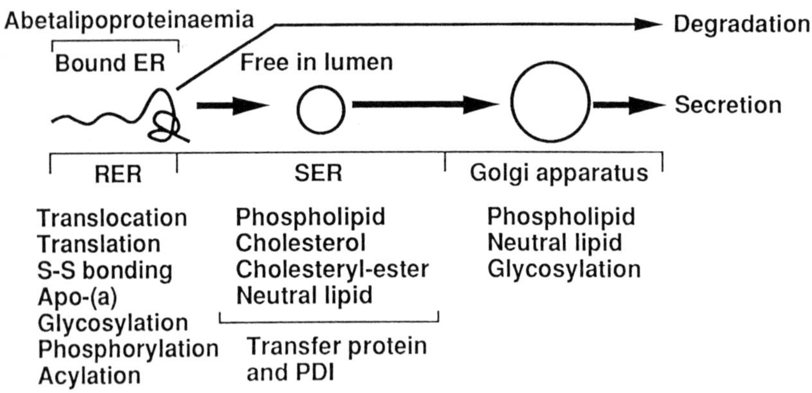

Figure. 3 Diagrammatic representation of lipoprotein assembly. Apo B100 is co-translationally inserted into the inner leaflet of the rough endoplasmic reticulum (RER). From this site it buds into the lumen, having received lipid from its sites of synthesis in the outer leaflet of the ER. The effect in abetalipoproteinaemia is predicted to act at this stage. After this, the nascent lipoprotein can undergo the distinct fates of intracellular degradation or secretion. Lipid availability mediates this event. PDI, protein disulphide isomerase.

translation and translocation studies in our laboratory indicate that nascent apo B is co-translationally inserted into the inner leaflet of the ER from which it must bud to form nascent lipoprotein particles (Pease *et al.*, 1991). After budding, these particles are capable of secretion (Borén *et al.*, 1992).

Abetalipoproteinaemia is a Mendelian recessive disorder which, in homozygotes, leads to the complete absence of apo B-containing lipoproteins in the circulation. Genetic studies have shown that this disorder is not a defect of the apo B gene; these studies, in conjunction with immunocytochemical studies and lipoprotein assembly studies, suggest that the defect in this disorder could be of a 'chaperone' protein that is necessary for apo B-containing lipoproteins to bud into the secretory pathways (Talmud *et al.*, 1988). The gene encoding a microsomal triglyceride transfer protein has been shown to be defective in three patients with classical abetalipoproteinaemia (Sharp *et al.*, 1993; Shoulders *et al.*, 1993). Homology searches have shown that this protein belongs to the vitellogenin gene family. The homologous region has been shown by X-ray crystallography to contain a lipid-binding cavity.

After its biosynthesis, apo B can undergo two distinct fates (Scott, 1990). Additional lipid may be added to the nascent lipoprotein in the ER (probably mediated by a neutral lipid transfer protein; Wetterau *et al.*, 1992) and Golgi apparatus and the lipoprotein particles secreted, or apo B, which is believed to have remained membrane-bound, is degraded in an ER-related compartment (White *et al.*, 1992). Thus, in an experimental cellular system, between 50 and 90% of newly synthesized apo B undergoes degradation in the ER.

4. The regulation of apo B-containing lipoprotein secretion

Dietary lipid intake and the metabolic abnormalities that occur in diabetes mellitus and chronic hypoalbuminaemic renal disease and in the genetic disorder familial combined hyperlipidaemia are all considered to produce elevated levels of apo B-containing lipoprotein particles from the liver, and this partly accounts for the hyperlipidaemia which occurs in these conditions. Thus hyperlipidaemia is an issue of considerable importance in understanding the regulation of the production of apo B-containing lipoproteins. Short-term studies in tissue culture cells indicate that the stimulation of apo B-containing lipoprotein secretion by lipid availability and its inhibition by insulin or by albumin in the environment of the tissue culture cells is not mediated by alteration at the apo B mRNA level, but rather by co- or post-translational mechanisms (*Figure 4*). *A priori*, this regulation could be mediated at the level of translation of apo B mRNA, or at the level of intracellular degradation versus secretion, or through modulation of a stored pool of apo B-containing lipoprotein. As there is no evidence to suggest the existence of a pre-secretory stored pool of lipoprotein particles, we have chosen to distinguish the other two possible mechanisms of modulation of apo B-containing lipoprotein secretion by indirect methods.

To achieve this goal, stable transformants of the rat hepatoma McArdle 7777, containing carboxy-terminally truncated forms of apo B between apo B13 and apo B 41, were used (White *et al.*, 1992). The constructs expressing the truncated forms were driven from the SV40 promoter and utilized the SV40 polyadenylation signals (*Figure 5*). The 5′-untranslated region of the constructs was derived from apo B mRNA. Thus any differences in the behaviour of the constructs should depend on the extent of the carboxy-terminal coding sequence of the protein.

The constructs apo B41, apo B36, apo B32 and apo B29 all responded to oleate with increased secretion; apo B23 showed a variable response; apo B17 and apo B13 showed no stimulation. The density of the truncated forms and of endogenous rat apo B100 and B48 was examined in continuous caesium chloride density gradient. Apo B100 was secreted on a particle with the density of VLDL; apo B48, apo B41, apo B36, apo B32, apo B29, and apo B23 to a variable extent, were secreted on a particle with the density of HDL; apo B17 and apo B13 were secreted with the bulk of the soluble protein and the density of albumin.

We conclude from this series of experiments that the ability of apo B

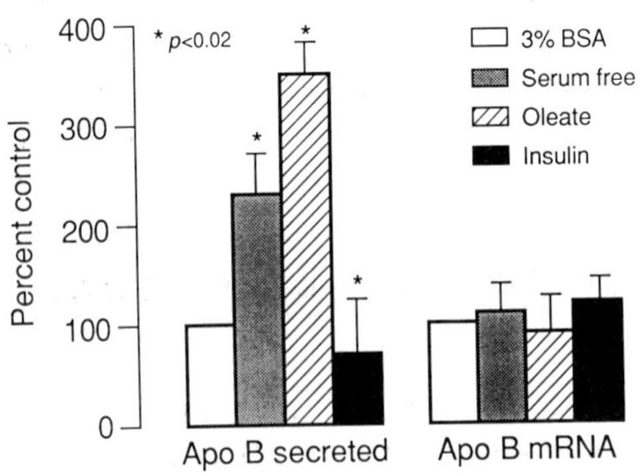

Figure 4. Regulation of secreted apo B and of apo B mRNA levels.

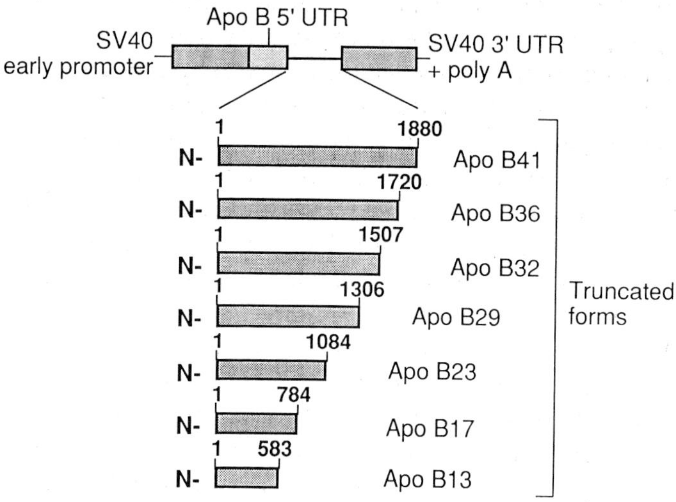

Figure 5. Carboxy-terminally truncated constructs of apo B. UTR, untranslated region.

to respond to oleate is a function of lipoprotein assembly. It is unlikely to be due to increased translation of apo B mRNA because all of the constructs contain the same untranslated regions. Therefore, the ability of the constructs to respond to oleate is most likely to be due to a decrease in pre-secretory degradation as a consequence of increased lipoprotein formation. Those constructs that do not direct synthesis of lipoprotein particles show no response to oleate.

For the above interpretation to be correct, it is necessary that the truncated constructs are indeed subject to pre-secretory degradation. Therefore, the extent of apo B secretion compared to the amount synthesized was compared for the various constructs. Each of the truncated forms showed non-quantitative secretion in the absence of intracellular retention, thus confirming that they undergo pre-secretory degradation (*Figure 6*).

5. The editing of apo B mRNA and the biosynthesis of apo B48

There is a single copy of the human apo B gene, spanning 43 kb and divided into 29 exons (Hodges and Scott, 1992; Scott, 1989a,b, 1990). A 14 kb mRNA encodes the 512 kDa apo B100 protein. In the middle of the huge (7572 base) exon 26, RNA editing of the mRNA or pre mRNA

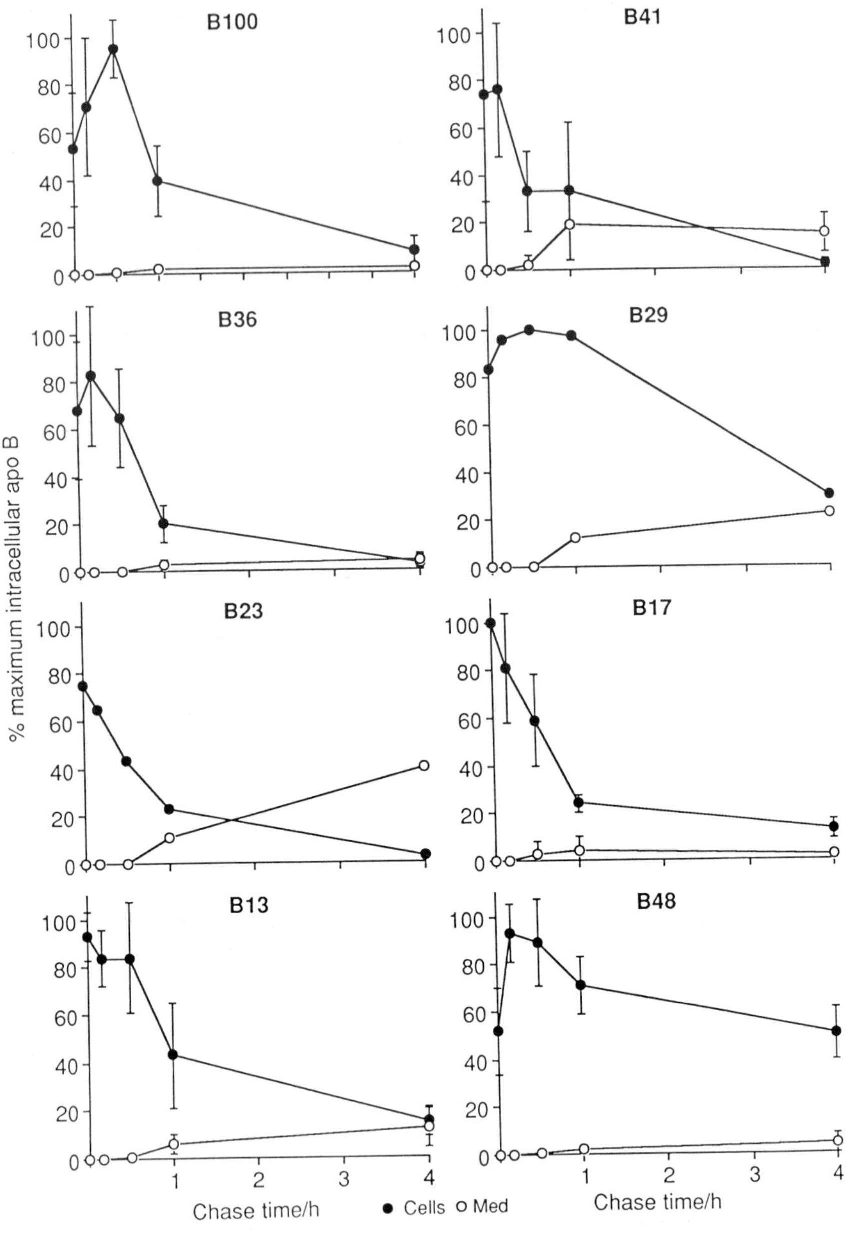

Figure 6. Pulse-chase studies of intracellular and secreted apo B constructs.

changes codon 2153 (CAA, glutamine) to a stop translation codon (UAA) (*Figure 7*). Some of the edited RNA is subsequently truncated at cryptic polyadenylation sites to create 7 kb mRNAs. Apo B48 protein (241 kDa) consists of the amino-terminal 2152 amino acids of apo B100. It is fully competent for lipoprotein assembly and secretion, but lacks the carboxy-terminal domain of apo B100 that mediates LDL receptor binding.

5.1. Biochemical characteristics of the apo B-RNA-editing enzyme

A breakthrough in the investigation of apo B-mRNA editing was the development of an *in vitro* reaction that edits exogenously added apo B RNA and a simple, sensitive primer extension assay. The primary source of editing activity has been cytoplasmic S100 extracts, but the cellular localization of the editing activity is uncertain. Apo B mRNA can be edited post-transcriptionally; nuclear extracts are functional, nuclei are enriched for unedited transcripts (D.M. Driscoll and J. Scott, unpublished data) and editing can influence 3′ end formation and polyadenylation. These results are indicative of a post-transcriptional nuclear function, possibly operating late in the pathway of mRNA maturation.

The editing activity of enterocyte extracts has been partially purified and initial biochemical characterization carried out. The reaction shows an apparent K_m for the RNA substrate of approximately 2 M and a V_{max} of 10^{-16} mol^{-1} μg^{-1} crude extract protein. The editing extracts are active

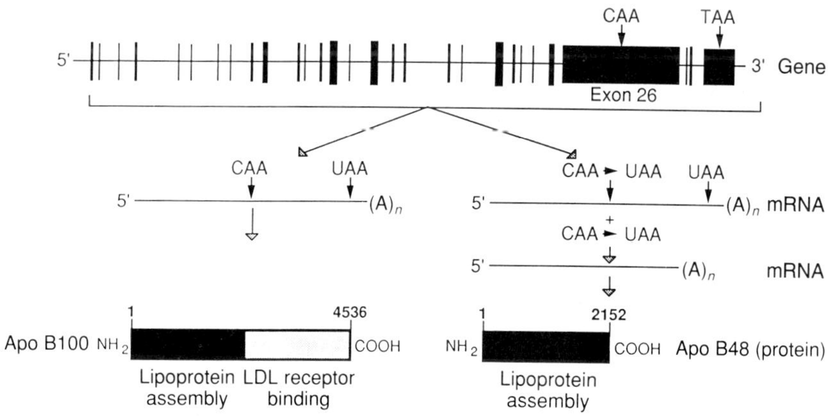

Figure 7. The intron–exon organization of the apo B gene and origin of apo B100 and apo B48 mRNAs.

after dialysis, ammonium sulphate precipitation, ion exchange chromatography and gel filtration, indicating no requirement for low molecular weight co-factors or a second substrate. The reaction does not require an added energy source or divalent cations, although crude preparations are stimulated by chelators. The activity is inhibited by high salt (>200 mM) and by vanadyl ribonucleoside complexes (>1 mM), but not by tetrahydrouridine, an inhibitor of cytidine deaminase.

The apo B editing enzyme is a protein or protein complex with no detectable RNA component. The activity is destroyed by proteases, heat denaturation, and by sulphydryl, imidazole or guanidino modification, and has the buoyant density of pure protein (1.3 g ml^{-1}). While the editing activity of a crude extract can be inhibited by nucleases, partially purified editing enzyme is no longer sensitive, suggesting that nuclease treatment generates oligoribonucleotide inhibitors of the editing enzyme. Its size by gel filtration has been estimated as 50–125 kDa; it sediments through glycerol gradients at 11S (240 kDa) (Hodges and Scott, 1992; Navaratnam *et al.*, 1993).

In summary, therefore, the apo B-mRNA editing activity consists of a well-defined enzyme with no RNA component. The editing enzyme is most likely to be a site-specific cytidine deaminase and bears little resemblance to complex mRNA processing events such as splicing and, unlike editing in kinetoplastid protozoa, does not utilize guide RNAs.

5.2. Mechanism of apo B-mRNA editing

When [^{32}P]cytidine triphosphate is incorporated into an apo B-mRNA substrate, *in vitro* editing specifically creates a [^{32}P]uridine monophosphate residue solely at the editing site (Hodges and Scott, 1992). Several important conclusions can be drawn from this simple observation. First, the product of the RNA editing is a simple uridine base; no complex, hypermodified base is created. Secondly, the 5′ phosphate of the base is preserved. There is no known polymerase that can remove a nucleotide and add a new nucleotide in its place without replacing the 5′ phosphate with a donor phosphate; so, the editing reaction does not involve excision and replacement. Thirdly, the process of RNA editing most probably involves site-specific cytidine deamination in the RNA. An alternative is transglycosylation, whereby the phosphoribosyl chain is maintained and a new base is exchanged into the ribose. However, no donor base has been identified, or is lost by filtration. Finally, the reaction could be a transamination, capturing the amine to transfer it to another molecule or perhaps catalysing the forward and reverse reactions interconverting U and C residues.

5.3. The sequence requirements for apo B-mRNA editing

Previously, we and others have demonstrated, by the transfection of apo B minigenes into cells that edit apo B mRNA, that between 354 and 26 nucleotides of apo B mRNA are sufficient to direct efficient editing, and that, using the *in vitro* system, 55 nucleotides and more direct editing.

Using deletion constructs, we have now shown that 42 nucleotides of apo B mRNA and 26 nucleotides (positions) of apo B mRNA direct editing with 50 and 25%, respectively, editing efficiency of the 55-base construct. We therefore decided to use the 55-base construct as the basis for construction of a series of 6-nucleotide scanning mutants in which the apo B sequence was changed to its complement (*Figure 8*; Yao *et al.*, 1991). The scanning mutants affecting the 12 nucleotides immediately downstream of the editing site completely abolished editing. The mutants immediately upstream and downstream of this 12-nucleotide segment reduced editing by approximately 50%. A series of point mutations was therefore constructed within this segment. This led to the identification of an 11-nucleotide sequence downstream of the edited C (GAUCAGUAU; position +5 to +13) where most mutations abolished or greatly decreased editing *in vitro*. In agreement with the

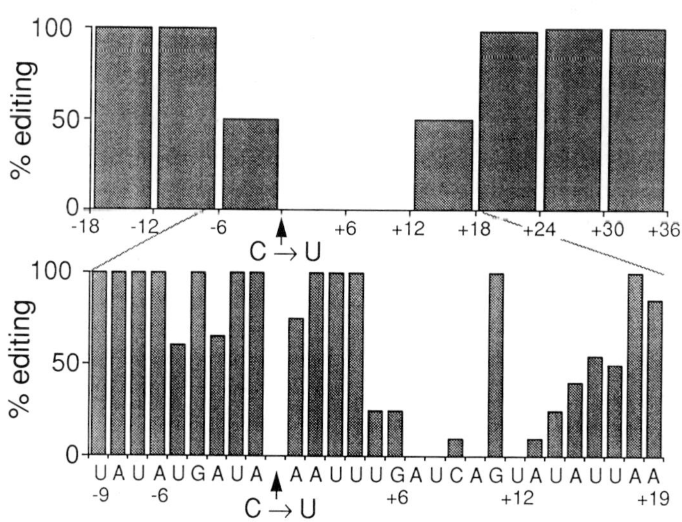

Figure 8. Mutational analysis of the apo B-mRNA editing site.

work of others, mutation in other positions proximate to the editing site led to little change or even to increased editing. These studies, together with the identification of a second apo B-mRNA editing site affecting nucleotide C-6802, indicate that a downstream sequence is necessary for editing to occur (*Figure 9*).

The interaction of cellular proteins with apo B mRNA has been investigated by glycerol gradient sedimentation, RNA gel mobility shift and RNA–protein cross-linking by UV light. Smith and co-workers (Harris *et al.*, 1993) have demonstrated, by gradient sedimentation, that the apo B-RNA substrate assembles with rat liver proteins over 3 h into 11S (240 kDa) and then 27S (1400 kDa) complexes, and they have proposed that the latter is the active 'editosome' – after the term spliceosome (Hodges and Scott, 1992). We have also demonstrated that, before binding RNA, the editing activity from rat enterocytes sediments at 11S (Hodges and Scott, 1992). However, in this study no lag period was required prior to catalysis and no higher order complex was assembled on the RNA substrate. Hence, RNA editing activity showed no similarity to the spliceosomal or polyadenylation complexes. The RNA mobility shift assay shows that proteins bind specifically to apo B RNA. UV light cross-links a 60 kDa protein to apo B mRNA, with specificity for the editing site. Editing activity directly correlates with this in inhibitor studies, partial purification, and studies with mutant RNAs that edit and do not edit (*Figure 10*). The sequence of p60 indicates that it is homologous to proteins containing an RNA recognition motif (*Figure 11*). The catalytic subunit of the apoB-mRNA editing enzyme has recently been identified by expression cloning and shown to encode a member of the cytidine deaminase family with strong homologies to all other known cytidine deaminases (Navaratnam *et al.*, 1993; Teng *et al.*, 1993).

			C→U		****** ****
Human	*site 1*	GAUA	C	AAUU	UGAUCAGUAUAUUAAAG
Baboon	*site 1*	GAUA	C	AAUU	UGAUCAGUAUAUUAAAG
Pig	*site 1*	GAUA	C	AAUU	UGAUCAGUAUAUUAAAG
Rabbit	*site 1*	GAUA	C	AAUU	UGAUCAGUAUAUUAAAG
Rat	*site 1*	GAUA	C	AAUU	UGAUCAGUAUAUUAGAG
Mouse	*site 1*	GAUA	C	AAUU	UGAUCAGUAUAUUAAAG
Human	*site 2*	AAAA	C	AAUCCA	UGAUCUACAUUUGUUUA
Chick		GGUG	C	AAAU	UGAGCAGUACAUCAAAG
					****** ****

Figure 9. Comparison of the conserved sequence between species and between the primary and secondary apo B-mRNA editing sites.

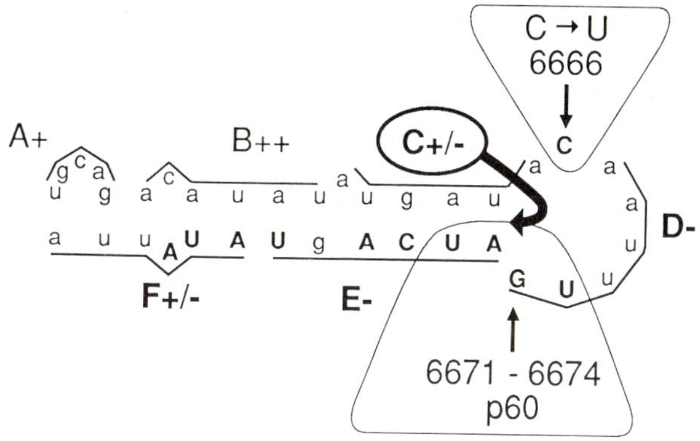

Figure 10. The apo B-mRNA editing site. A highly conserved stem-loop is predicted by the computer. The editing site and p60 binding site are indicated. The position of scanning mutants A–F are indicated. Mutants D and E abolish editing; mutants C and F reduce it by 50%. Mutant C would be predicted to disrupt the stem-loop. Bases shown in upper case are the editing site and the position of point mutations that abolish editing.

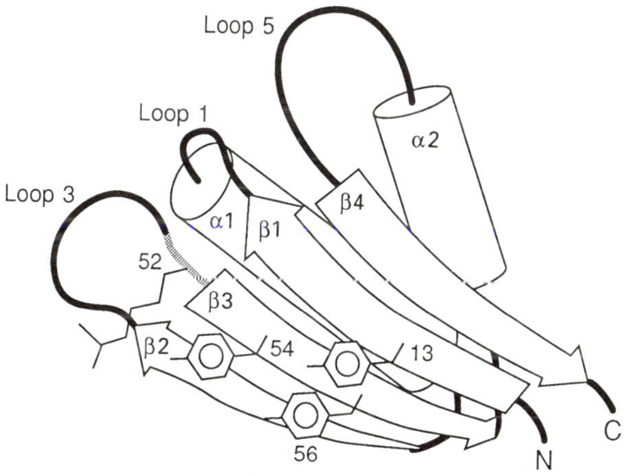

Figure 11. The NMR structure of an RNA recognition motif (RRM)-binding protein is delineated (Kenan *et al.*, 1991). The position of α-helices (α) and β-sheets (β) together with phenylalanine (amino acids 13, 54, 56) and arginine residues (amino acid 52) believed to be involved in RNA binding are delineated. Reproduced from Kenan *et al.* (1991) with permission from Elsevier Trends Journals.

6. Conclusions

The structure and functional domains of apo B100 have been defined. Throughout its length, apo B100 is a lipid-binding protein which, in its

3

Prospects for gene therapy of X-linked immunodeficiency diseases

C. Kinnon
Molecular Immunology Unit, Division of Cell and Molecular Biology, Institute of Child Health, London, UK

1. Introduction

There are seven well-defined X-linked immunodeficiency disorders (*Figure 1*). These diseases arise as consequences of mutations occurring naturally in the molecular components of the immune system. Until recently, the molecular bases of the X-linked immunodeficiency disorders were largely unknown. However, with the many recent advances in recombinant DNA technology, the defective genes in several of these diseases are now being elucidated (summarized in *Figure 1*).

The first of these genes to be identified was gp91-*phox* which, when defective, causes X-linked chronic granulomatous disease (CGD; Royer-Pokora *et al.*, 1986). This was also the first successful use of the positional cloning approach for human genetic diseases, a process which involves the detailed genetic and physical mapping of the gene locus (reviewed in Ballabio, 1993; Collins, 1992). Genetic linkage studies of affected families and the construction of long-range physical maps of the relevant chromosomal region enable reduction of the extent of DNA over which the gene can be contained. Once this is reduced to a manageable size, usually about 2 Mb or a size that is small enough to be contained in yeast artifical chromosomes (YACs) and cosmids, cDNA

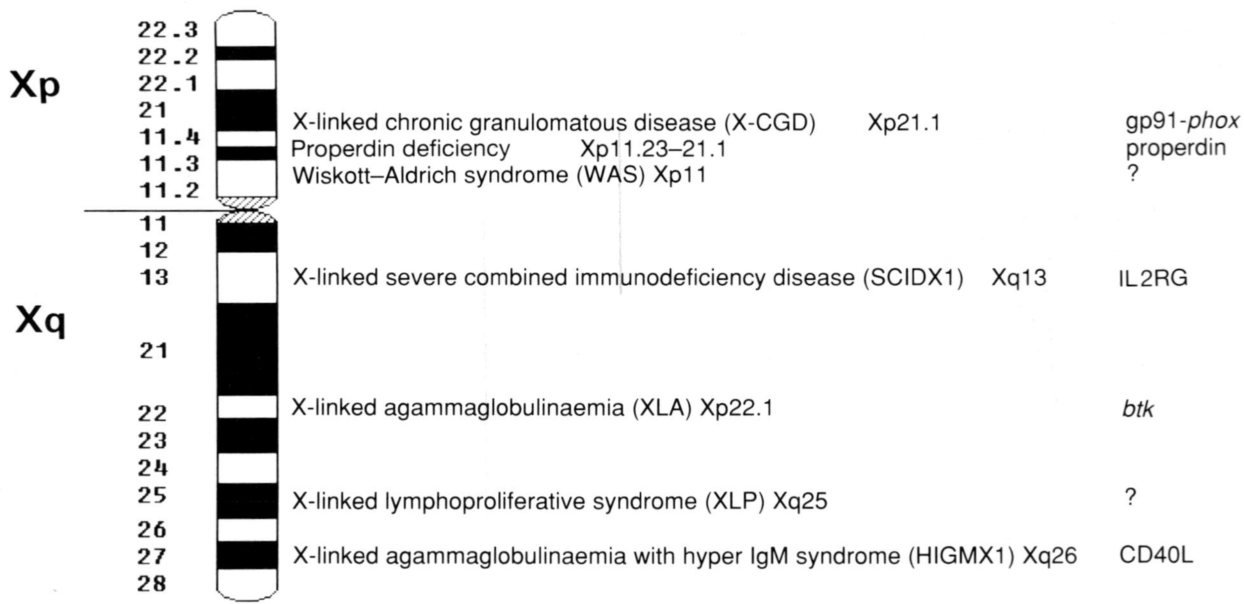

Figure 1. A schematic representation of the human X chromosome, showing the approximate positions of the immunodeficiency disease loci and the genes which cause these defects.

selection or exon trapping techniques can be used to identify potential coding sequences.

The success of this approach opened the way for the mapping of the other disease loci; in the first half of 1993 the identification of the genes responsible for three of these disorders, namely X-linked agammaglobulinaemia (XLA or AGMX1), X-linked agammaglobulinaemia with hyper IgM syndrome (HIGMX1) and X-linked severe combined immunodeficiency (SCIDX1), were reported. The gene responsible for XLA, *btk*, was found to encode a non-receptor tyrosine kinase expressed in B cells (Tsukada *et al.*, 1993; Vetrie *et al.*, 1993), the CD40 ligand (CD40L) was suggested as a candidate gene for HIGMX1 (Allen *et al.*, 1993a; Aruffo *et al.*, 1993; DiSanto *et al.*, 1993; Korthäuer *et al.*, 1993; Ramesh *et al.*, 1993) and, similarly, the γ-chain of the interleukin-2 receptor (IL-2RG) was suggested for SCIDX1 (Noguchi *et al.*, 1993b; Puck *et al.*, 1993b). As yet, the genes for Wiskott–Aldrich syndrome (WAS) and X-linked lymphoproliferative syndrome (XLP) have not been identified but, with the current rate of progress, it must be expected that their elucidation will not be very long coming.

The precise mapping of the genes causing the various immunodeficiency diseases led to opportunities for the prevention of these diseases through the identification of carrier females in affected families and prenatal diagnosis. This has been achieved through the use of assays based on X-inactivation analysis and gene tracking using closely linked probes with restriction fragment length polymorphisms (RFLPs). In the cases where the genes are now identified, it is possible to diagnose the diseases and assess carrier status for at risk females unambiguously, based on the identification of the mutations in affected families and through the use of RFLPs associated with the genes themselves. Furthermore, the identification of the causative genes should lead to improved understanding of the nature of these disorders and, in time, to their improved management and treatment. Perhaps the most exciting prospect in this area is treatment by somatic gene therapy. The development of suitable techniques is now underway for several of these disorders. Since it was the first of these disorders to be elucidated at a molecular level, we have so far concentrated on developing such techniques for treating CGD. It should be possible to extrapolate some of the results of this work to the other immunodeficiency disorders. Work by many groups in this area has already resulted in attempts to treat patients with an autosomal recessive form of severe combined immunodeficiency (SCID), caused by deficiency of the adenosine deaminase (ADA) enzyme, by somatic gene therapy.

In this chapter, the nature of some of these disorders is discussed together with what is currently known about the defective genes and/or their protein products. In addition, the possibilities for improved counselling and treatment, particularly somatic gene therapy, are also discussed.

2. X-linked agammaglobulinaemia

XLA is characterized by agammaglobulinaemia involving all immunoglobulin subclasses, and is the direct result of a lack of mature B cells. Pre-B cells are present at normal levels in the bone marrow and the defect appears to involve a block in their differentiation. The defect does not appear to affect the process of immunoglobulin gene rearrangement directly, since these rearrangements appear to occur normally, resulting in apparently functional antibodies, in the small numbers of B cells isolated and analysed from these patients.

The disorder is usually diagnosed quite early in life as a result of the number of episodes of severe bacterial infection in affected boys. Once diagnosed, the usual treatment is a lifetime of regular intravenous immunoglobulin infusions. This significantly reduces the episodes of infection and, with good management, the life expectancy of most boys with this condition is extended to a near normal length. However, the treatment is not a long-term cure and some of these patients succumb to the long-term consequences of bacterial infection and infection with rare neurotropic viruses.

The gene for this disorder was mapped to Xq22, more precisely to the region between the loci DXS442 and DXS101, a region of about 2–4 cM (Lovering *et al.*, 1993). No recombinations were observed between the DXS178 locus, which lies in the centre of this region, and the XLA locus in over 70 chromosomes investigated (Lovering *et al.*, 1993). A DXS178-positive YAC was screened for B-cell-specific cDNAs and one of these was found to encode a novel tyrosine kinase gene, *btk* (also known as *atk*; Vetrie *et al.*, 1993). The *btk* gene was found to be mutated in eight unrelated XLA patients, suggesting that defects in this gene cause XLA (Vetrie *et al.*, 1993). At the same time, a second group isolated a specific tyrosine kinase gene from mouse pre-B cells, which mapped to this region of the X chromosome (known as *bpk*; Tsukada *et al.*, 1993). Antisera to the protein product of this gene demonstrated reduced levels of protein and kinase activity in cell lines derived from XLA patients (Tsukada *et al.*, 1993). This gene is the mouse equivalent of *btk*, as demonstrated by 98% homology between their deduced amino acid sequences (Tsukada *et al.*, 1993; Vetrie *et al.*, 1993).

The predicted amino acid sequence of btk is highly homologous to members of the src-related non-receptor tyrosine kinase family, which includes many of the tyrosine kinases thought to be involved in signal transduction in haematopoietic cells (e.g. fyn, lck and lyn). Based on homology between members of this family and others, the btk protein sequence can be subdivided into four separate functional domains comprising a unique N-terminal domain, an src-homology-3 (SH3) domain, an src-homology-2 (SH2) domain and a tyrosine kinase catalytic domain (*Figure* 2). The SH2 and SH3 domains have been identified in a wide variety of different signal transduction molecules (reviewed in Pawson and Gish, 1992). Deletion analysis of src-related tyrosine kinases suggests that these domains may be involved in intermolecular recognition and regulation of kinase activity. SH2 domains have been shown to recognize phosphorylated tyrosine residues within a specific peptide sequence. X-ray crystallography and modelling of the src SH2 domain suggest that this domain forms a two-holed socket which can bind the specific phosphorylated residues (Waksman *et al.*, 1993). Binding of SH3 domains to specific peptide sequences has also been observed (Ren *et al.*, 1993).

The mutations in the *btk* gene detected so far appear to be quite varied (Bradley *et al.*, 1994; Lovering *et al.*, 1994; Vetrie *et al.*, 1993; de Weers *et al.*, 1994). The identification of deletions within the *btk* gene in some XLA patients (about 15%) indicates that the disorder is likely to arise as a result of a loss of function of the protein. The initial mutation analysis of the *btk* gene in XLA patients revealed several intragenic deletions and two point mutations resulting in amino acid substitutions, one of which is in the ATP-binding site of the catalytic domain, and so likely to reduce the kinase activity, and the other is in the

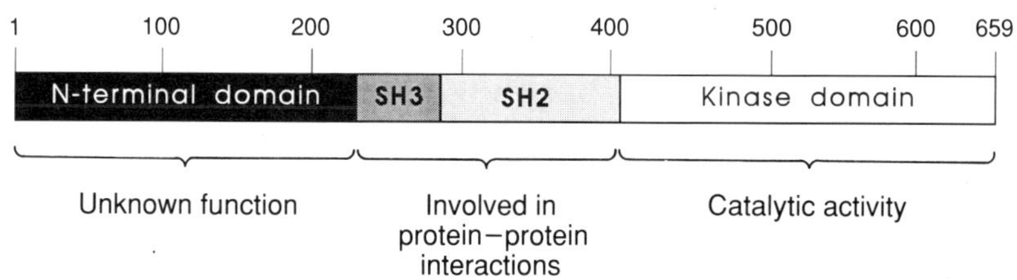

Figure 2. A schematic representation of the btk protein, highlighting the functional domains. An approximate scale is given in amino acid residues.

conserved substrate recognition sequence of the kinase domain (Vetrie *et al.*, 1993; *Table 1*). While only one additional gene deletion has been identified so far (Lovering *et al.*, 1994), a total of 12 additional point mutations have now been identified, including amino acid substitutions in the SH2 domain (Bradley *et al.*, 1994; de Weers *et al.*, 1994; summarized in *Table 1*). Two patients, suffering with XLA of markedly different severity, have been identified with amino acid substitutions at conserved arginine residues in the SH2 domain. One patient (JP; *Table 1*), with the classical XLA phenotype, is mutated at the arginine residue which is predicted to be directly involved in binding the phosphotyrosine residue at the base of the binding pocket, while the other, less severely affected, patient (276; *Table 1*) has a mutation in the arginine residue predicted to lie at the lip of the pocket and interact with the peptide backbone (Bradley *et al.*, 1994; Waksman *et al.*, 1993; de Weers *et al.*, 1994). Thus, although both mutations can be predicted to disrupt phosphotyrosine binding, the mutation at the phosphotyrosine binding site results in a more severely affected phenotype. Similar differences in severity of disease phenotype have been observed to correlate with different mutations in other parts of the *btk* gene (Bradley *et al.*, 1994; *Table 1*). Continued mutation analysis of both classical and atypical XLA patients will allow a greater characterization of the functionally important residues in the protein.

The identification of such loss-of-function mutants will provide a unique opportunity to study the B-cell differentiation pathway from within the cell and to relate structure to function in this protein. Such results potentially could be extrapolated to other members of the *src* gene family. In addition, *in vitro* experiments on the interaction of btk with other cellular components should shed light on the complex signalling requirements of developing B cells.

It is now possible to give more accurate diagnosis of XLA, based on deletion analysis and mutation detection. In families where gene deletions have been identified in affected boys, it will now be possible to give carrier status diagnoses to their female relatives with absolute certainty (Lovering *et al.*, 1994). As discussed, the majority of cases are due to point mutations, small insertions or deletions which can be detected by cDNA sequence analysis. Even though these patients do not have B cells, *btk* mRNA can be detected in their peripheral blood cells, as the gene is also expressed in myeloid cells (Vetrie *et al.*, 1993). In order to identify mutations quickly in these families, we have used single-strand conformation polymorphism (SSCP) analysis of the *btk* gene (Bradley *et al.*, 1994). Normal and target cDNA are amplified using

Table 1. Mutations in the *btk* gene

Patient	cDNA position[e]	Base change	Result
CB[a]	133	A→G	Met (1), start codon mutation
BB[a]	134	T→C	Met (1), start codon mutation
M76[b]	215	G→A	Arg (28) → His in unique N region
JG[a]	442	21 bp insertion	Seven amino acid insertion (103) in unique N region
B225[b]	721	8 bp insertion	Frameshift at Ile (197), premature stop (201)
NAD[a]	895	C→T	Premature stop (255) in SH3 domain
276[a,b,d]	994	C→T	Arg (288) → Trp in SH2 domain
JP[a]	1051	A→G	Arg (307) → Gly in SH2 domain
JL6[b]	1407	C→A	Premature stop (425) in ATP-binding site
B[c]	1420	A→G	Lys (430) → Glu in the ATP-binding site
A[c]	1706	G→A	Arg (525) → Gln in substrate-specific region
JF[a,d]	1952	C→A	Ala (607) → Asp near 3′ end
AP[a]	2038	G→T	Premature stop (636), loss of last 24 residues
G03[b]	2041	6 bp insertion	Premature stop (638), loss of last 22 residues

Data from: [a] Bradley *et al.* (1994); [b] de Weers *et al.* (1994); [c] Vetrie *et al.* (1993); [d] Indicates patients with a less severe 'leaky' phenotype; [e] cDNA numbering as in Vetrie *et al.* (1993).

polymerase chain reaction (PCR) amplification, denatured to single-stranded DNA and then the single strands are separated by non-denaturing polyacrylamide gel electrophoresis. The two single-stranded DNA molecules assume a conformation which is sequence dependent, a single base pair change results in an altered conformation and a shift in mobility of the DNA in this analysis. SSCP is a simple and sensitive technique for the relatively rapid screening of patient samples. Use of this technique for a given gene usually detects 70–95% of point mutations (reviewed in Grompe, 1993). PCR amplification of limited regions of the *btk* cDNA, followed by SSCP analysis, allows mutations to be identified in these amplified fragments. Only the fragments containing the mutations require sequencing, so the amount of sequence analysis necessary is reduced substantially.

Where the mutations have been identified, it is now possible to identify female relatives of affected boys who may be carriers. In cases where the mutation results in the introduction or deletion of a restriction site, the analysis can be performed by a simple restriction digest of the cDNA derived from peripheral blood (Vetrie *et al.*, 1993). In the remainder of cases, the majority, it will be necessary to visualize the mutation by SSCP analysis. It is possible to visualize the mutations by SSCP analysis of cDNA prepared from the mononuclear cells of whole blood samples of obligate carriers of XLA, where point mutations have already been identified (our unpublished results). We have also identified a polymorphism in the *btk* gene using SSCP analysis which represents a base change at the nucleotide level and does not result in an RFLP or a change in the amino acid sequence (Bradley *et al.*, 1994; Katz *et al.*, 1994). As the heterozygosity of this polymorphism is about 35%, it may be useful for carrier determination in some families.

With an improved understanding of the role of btk in XLA and normal immune function it will be possible to develop new treatments, including the possibility of somatic gene therapy. From what we know already it is likely that reconstitution of btk activity, even at low levels, in B cells at the correct stage of development should be sufficient to confer a growth advantage to those cells. However, before such a treatment can be considered, it will be important to investigate the effects of over-expression of btk, as it is possible that it may have a potential for oncogenesis. As discussed, XLA results from loss of function of btk, in contrast to diseases where constitutive activation of the kinase domain is associated with an oncogenic potential (e.g. v-src and BCR-abl). We have evidence to suggest that btk probably does not play a crucial role in at least one form of B-cell leukaemia, B-cell acute lymphoblastic leukaemia (ALL). On investigating leukaemic cells from 23 such patients, btk was found to be expressed at levels corresponding to those seen in unaffected individuals with no evidence of mutations in the gene (Katz *et al.*, 1994). The generation of a transgenic mouse over-expressing the btk protein will provide more definitive evidence for a lack of oncogenic potential for btk. Reconstitution of a complete immune system in an immunodeficient mouse with a mutation in btk, the *xid* mouse strain (Rawlings *et al.*, 1993; Thomas *et al.*, 1993), would be a useful first step in the development of such techniques for eventual clinical use.

3. X-linked agammaglobulinaemia with hyper IgM syndrome

HIGMX1 is characterized by an agammaglobulinaemia which is distinct from XLA in that only the antibodies of the IgG and IgA subclasses are absent and there are normal or aberrantly high levels of IgM antibodies. The IgM antibodies may, in some cases, be monomeric and may lack function. B cells are present at normal levels in this disease. X-inactivation studies in obligate female carriers of HIGMX1 showed that there is random inactivation in the T- and B-cell lineages, which supports the idea that the class switch mechanism in B cells is intact (Hendriks *et al.*, 1990).

Although the diagnosis of this disease is relatively straightforward, some cases are misdiagnosed as XLA. In addition to the elevated levels of IgM, affected boys are more prone to infections more typical of T-cell deficiencies, such as *Pneumocystis carinii* pneumonia and cryptosporidial diarrhoea. As with XLA, the disorder is treated by regular immunoglobulin infusion.

The HIGMX1 locus had only been mapped crudely to Xq24–26, largely because of the rarity of families suffering from this disorder; however in 1992 it was shown to map close to the HPRT locus in Xq26 (Padayachee *et al.*, 1992). At around the same time, the gene for CD40L was isolated in several laboratories (Graf *et al.*, 1992; Hollenbaugh *et al.*, 1992; Spriggs *et al.*, 1992). CD40L is expressed on activated T cells, binds the CD40 molecule expressed on B cells and stimulates their proliferation. Co-stimulation of B cells with CD40L and interleukins IL-2 or IL-4 induces IgG or IgE secretion, respectively. Taken together with the knowledge that CD40L maps to Xq26, near to the HPRT locus, this made it a very good candidate for the gene which is defective in HIGMX1. CD40L is a type II integral membrane glycoprotein expressed on activated T cells. It has some sequence homology to the tumour necrosis factor (TNF) and a predicted tertiary structure similar to the TNF trimer.

Five groups have now analysed a total of 14 HIGMX1 patients from different families for mutations in their CD40L genes (Allen *et al.*, 1993a; Aruffo *et al.*, 1993; DiSanto *et al.*, 1993; Korthäuer *et al.*, 1993; Ramesh *et al.*, 1993; summarized in *Table 2*). As with XLA, these mutations are varied and may provide useful insights into the function of this protein. They are unique to individual families and will provide a basis for carrier determination and prenatal diagnosis. A dinucleotide repeat polymorphism has been identified in the 3′-untranslated region of the CD40L gene (Allen *et al.*, 1993b; DiSanto *et al.*, 1993). The heterozygosity, reported as being between 70 and 80%, depending on the ethnic origin of the population studied, suggests that this should be useful in ascertaining carrier status (Allen *et al.*, 1993b).

It has already been shown that CD40L can be expressed in transfected human B cell lines and that such cells can be used to induce isotype switching to IgE production in the presence of IL-4 (Allen *et al.*, 1993a). The expression of the normal CD40L gene in peripheral T cells from patients and the demonstration of their ability to induce switch recombination in B cells must be the next steps towards the develop-

Table 2. Mutations in the CD40L gene

Patient	cDNA position[f]	Base change	Result
AT[a]	109	T→G	Met (36) to Arg in transmembrane region
2[b]	289	58 bp deletion	Frameshift at Asp (96), premature stop (107)
P1[c]	348	63 bp deletion	21 residues deleted (116–136)
P3[c]	370	C→A	Ala (123) → Glu
CD[d]	386	T→A	Ser (128) → Arg
	388	A→G	Glu (129) → Gly
P4[c]	413	10 bp deletion	Splice site mutation (?) resulting in premature stop (144)
P2[c]	415	8 bp deletion	Splice site mutation (?) resulting in premature stop (138)
BW[a]	420	T→G	Trp (140) → Gly
TG[a]	421	G→A	Premature stop (140)
2[e]	–	–	Leu (155) → Pro
1[b]	590	C→A	Frameshift at Ser (197), premature stop (241)
	591	C deleted	
3[e]	–	–	Thr (211) → Asp
1[e]	–	–	Gly (227) → Val
JW[d]	705	G→C	Ala (235) → Pro

Data from: [a] Korthäuer *et al.* (1993); [b] Ramesh *et al.* (1993); [c] DiSanto *et al.* (1993); [d] Aruffo *et al.* (1993); [e] Allen *et al.* (1993a); [f] cDNA numbering from initiation codon, Met (1).

ment of somatic gene therapy for HIGMX1. The main problem that can be envisaged is that CD40L is usually only expressed in activated T cells and deregulated expression may lead to as yet unforeseen problems. Some of these problems could be overcome by placing the expression under the control of endogenous promoter/control elements.

4. X-linked severe combined immunodeficiency

The SCIDs comprise a syndrome that is characterized by absent cell-mediated and humoral immunity. All forms of SCID are generally diagnosed in early life and are usually fatal by 1–2 years of life if not successfully treated by bone marrow transplantation. SCID is generally treated by allogeneic bone marrow transplantation, with a 90% success rate using matched donors and a 60–70% success rate using mismatched donors. Since, in the majority of cases, the donors are mismatched, there is still a need for the development of more effective therapies.

The X-linked form of the disease, comprising about two-thirds of all cases, has recently been shown to be caused by defects in the IL-2 receptor (IL-2R) γ-chain gene (Noguchi *et al.*, 1993b; Puck *et al.*, 1993b). The IL-2RG gene was found to map to Xq13 (Noguchi *et al.*, 1993b; Puck *et al.*, 1993b), the same region as the SCIDX1 locus (Markiewicz *et al.*, 1993; Puck *et al.*, 1993a). So far, a total of seven unrelated SCIDX1 patients have been reported to have different mutations in the IL-2RG gene (Noguchi *et al.*, 1993b; Puck *et al.*, 1993b) and we are in the process of characterizing this gene in an additional five SCIDX1 patients (Clark *et al.*, unpublished observations). All the mutations are summarized in *Table 3*. Once again, the identification of mutations in the gene will have important implications for the families involved.

The IL-2R γ-chain is a transmembrane glycoprotein with a number of structural motifs, including four conserved cysteine residues and a WSXWS motif, characteristic of cytokine receptor superfamily members, a leucine zipper-like motif in the extracellular domain and an SH2-like domain in the cytoplasmic tail (*Figure 3*). Originally identified as a component of the high and intermediate affinity IL-2R that is required to achieve full ligand-binding affinity and internalization, the γ-chain is constitutively expressed in haematolymphoid cells and is now known to be a component of additional cytokine receptors, including the IL-4 and IL-7 receptors (IL-4R and IL-7R; Kondo *et al.*, 1993; Noguchi *et al.*, 1993a; Russell *et al.*, 1993). It is postulated that, in each of these receptors, the unique chain recognizes the specific

Table 3. Mutations in the IL-2RG gene

Patient	cDNA position[d]	Base change	Result
1[a]	186, exon 2	T→A	Premature stop at conserved Cys (62)
7[a]	341, exon 3	G→A	Gly (114) → Asp, next to conserved Cys
1[b]	355, exon 3	A→T	Premature stop (119), loss of 251 residues
5[a]	454, exon 3/intron 3	GT→AT	Disrupts splice donor site, premature stop (152)
6[a]	458, exon 4	T→A	Ile (153) → Asn, near to leucine zipper region
REY[c]	548, exon 4	T→C	Leu (183) → Ser, within leucine zipper region
AZZ[c]	595, exon 4/intron 4	G→A	Disrupts splice donor site, premature stop (199)
KHA[c]	671–686, exon 5	16 bp deletion	Frameshift (224), premature stop (267) and loss of WSXWS motif
WEB[c]	703, exon 5	C→T	Premature stop (235), before WSXWS motif
2[b] and MOR[c]	865, exon 7	C→T	Premature stop (289) in cytoplasmic domain, loss of last 81 residues
3[b]	923, exon 7	C→A	Premature stop (308) in cytoplasmic domain, loss of last 62 residues

[a] Puck *et al.* (1993b); [b] Noguchi *et al.* (1993b); [c] Clark *et al.*, unpublished observations; [d] cDNA numbering from initiation codon, Met (1).

cytokine, and thus confers the signalling specificity, while the γ-chain transduces the signal via its cytoplasmic domain. All three receptors are involved at different stages of T-cell and B-cell growth and maturation and the combined effect of disrupting their expression may augment the immune defect. This would explain the apparent discrepancy between the much milder phenotypes observed in IL-2-deficient humans and IL-2- and IL-4-deficient mice (reviewed in Kondo *et al.*, 1993; Noguchi *et al.*, 1993a; Russell *et al.*, 1993).

In the majority of SCIDX1 families, the nature of the presumed point mutations has yet to be identified. This means that, at present, it is still difficult to diagnose the X-linked form of this disease unambiguously,

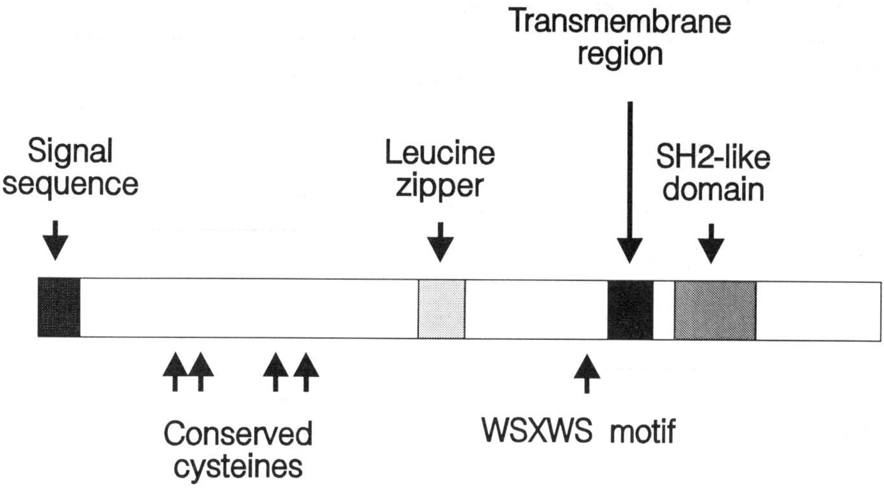

Figure 3. A schematic representation of the IL-2R γ-chain, highlighting the functional domains.

and to differentiate it from one of the autosomal recessive forms. This is particularly important in families where there is only one affected boy or no DNA samples available for mutation analysis. It is essential to ascertain whether the disorder is X-linked before genetic counselling can be provided based on associated polymorphisms either within, or closely associated with, the gene. The most effective way to determine X linkage is through X-inactivation analysis of purified T cells from female relatives of affected boys (Goodship *et al.*, 1989). This can provide about a 90% accuracy of diagnosis in about 90% of cases, using highly polymorphic dinucleotide repeat locus-specific probes (Goodship *et al.*, 1991).

No RFLPs have been found to be associated with the IL-2RG cDNA, however SSCP polymorphisms have been detected within both introns 1 and 2 of the gene, which may aid in carrier assessment and prenatal diagnosis in some families (Noguchi *et al.*, 1993b). In addition, the recent characterization of multiple, highly polymorphic dinucleotide repeat markers which flank the SCIDX1 locus means that genetic analysis can be performed in most families who have been diagnosed as SCIDX1 with over 99% accuracy (Markiewicz *et al.*, 1993).

SCIDX1 will undoubtedly be one of the prime target disorders for treatment by somatic gene therapy techniques. Indeed, there are several reasons why it might be expected that this molecule will become the main target for gene therapy protocols in the future. Not only do these include the severity of the disease and the relative abundance of patients, but also the protein is expressed constitutively in lymphoid cells, suggesting that problems of inappropriate tissue expression may be easier to overcome for this molecule. The fact that the IL-2R, IL-4R and IL-7R are widely studied cell surface molecules means that a great deal is already known concerning the biology of these molecules (reviewed in Miyajima *et al.*, 1992; Taniguchi and Minami, 1993). One form of SCID has already been treated by somatic gene therapy with some measure of success (see Section 6).

5. Chronic granulomatous disease

CGD is a group of related disorders characterized by defective phagocyte function. Phagocytic cells in children with CGD ingest organisms normally but fail to kill them as a result of defects in the superoxide-producing pathway. This results in recurrent and chronic bacterial and fungal infections in affected individuals. Treatment with prophylactic antibiotics, especially trimethoprim and sulfamethoxale, may partially control the frequency of infections. All forms of the disease affect the various components of the NADPH oxidase system of phagocytic cells (*Figure 4*). The NADPH oxidase is now known to consist of four specific components, two of which are cytosolic, p47-*phox* and p67-*phox*, and two are the membrane-bound flavocytochrome b_{558} components, gp91-*phox* and p22-*phox*. Defects in any of these components give rise to CGD, with over 90% of all cases caused by defects in gp91-*phox* or p47-*phox*.

Genetic analysis can be used for unambiguous carrier determination in about 60% of families affected by the X-linked form of this disease. A small percentage (about 15%) of patients have deletions within or involving the gp91-*phox* gene, which can be detected by cytogenetic analysis or altered patterns on Southern blot analysis. Although the point mutations in many families have now been identified (reviewed in Casimir and Teahan, 1994), carrier determination and prenatal diagnosis have relied largely on the identification of RFLPs. There are two intragenic RFLPs which should be informative in about 45% of families, allowing reliable carrier detection and first-trimester prenatal diagnosis (Pelham *et al.*, 1990). The majority (over 85%) of all families with defects in p47-*phox* could be diagnosed by virtue of an identical gene abnormality consisting of a dinucleotide deletion at a GTGT tandem repeat at the acceptor site of the first intron–exon boundary

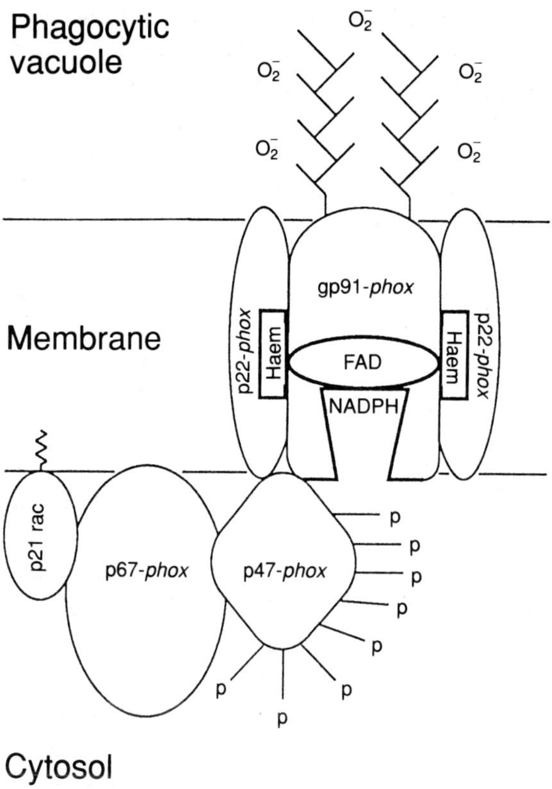

Figure 4. A schematic representation of the activated NADPH oxidase of neutrophils showing the stoichiometry of the complex, the components defective in CGD, p22-*phox*, p47-*phox*, p67-*phox* and gp91-*phox*, p21 rac and the haem, FAD and NADPH binding sites. Adapted from Casimir and Teahan (1994) with permission from Academic Press Inc.

(Casimir *et al.*, 1991). In families where genetic analysis is not possible, carrier detection can be performed using the nitroblue tetrazolium (NBT) test, with more accurate diagnosis depending on investigation of the presence or absence of relevant proteins in the neutrophils. Prenatal diagnosis can be performed using the NBT test on fetal blood samples.

Gene therapy experimentation in this area, to date, has made extensive use of retroviral vectors to introduce DNA sequences into the target cell population. As discussed elsewhere in this volume, this strategy utilizes the natural ability of retroviruses to transfer their nucleic acid content into a cell with high efficiency as these viruses efficiently integrate a DNA copy of their genome into the host chromatin. Such vectors are by far the most proven and developed vehicles used to date for therapeutic gene transfer. Retroviral vectors containing full-length cDNAs for both gp91-*phox* (Porter *et al.*, 1993) and p47-*phox* (Thrasher *et al.*, 1992) have been employed to transfect CGD patient lymphoid cell lines and demonstrated significant functional correction of the CGD defects in these lines. Such experiments must be viewed as the first steps in the process of developing the tools fcr the treatment of CGD by somatic gene therapy.

6. Future prospects

The identification of the causative genes in X-linked immunodeficiency disorders will lead to an improved understanding of the nature of these disorders and the potential for their more successful management and treatment. Immunodeficiency disorders are particularly good candidate diseases for treatment by somatic gene therapy. Since they are disorders of the immune system, the primary tissues to be treated, the peripheral blood cells, are readily accessible for genetic manipulation *ex vivo*. In addition, all the disorders discussed are curable by bone marrow transplantation. This means that the bone marrow can be used to provide an accessible source of stem cells. If self-renewing pluripotential stem cells could be transduced then, theoretically, it should be possible to treat a given patient once in a lifetime to correct the defect permanently. Transduction of the peripheral blood cells would mean repeated treatments at 2–6 month intervals throughout a patient's lifetime.

Several ADA–SCID patients have now been treated at different centres in the United States and Europe with retroviruses containing a functional copy of the ADA gene. The aim is to correct their defects by reconstituting expression in peripheral T cells. This has been attempted directly by introducing the gene into peripheral T cells expanded with IL-2 *in vitro*, so-called TIL cells. 'Stem' cell or progenitor cell populations have also been used, derived either from the peripheral blood or bone marrow. We have treated one patient with ADA–SCID by 'stem' cell somatic gene therapy. A population of bone marrow cells, containing stem and progenitor cell populations, was positively selected on the basis of CD34 expression by virtue of binding the ICH3 monoclonal antibody (Lebkowski *et al.*, 1992). This population of cells was transduced by co-cultivation with the POC-1 cell line, a retroviral producer cell line designed to introduce a functional copy of the ADA gene (van Beusechem *et al.*, 1990). Six months after gene transduction, the exogenous ADA gene can be detected in a random sample of bone marrow by PCR analysis (D. Valerio, P. Hoogerbrugge *et al.*, unpublished observations). The next step will be to investigate gene expression in this patient. So far the results of all these trials have not been published but anecdotal evidence suggests cautious optimism for the future of this field.

Acknowledgements

I would like to thank colleagues and members of my laboratory for helpful discussions and permission to cite unpublished data, particularly Ms Linda Bradley, Dr Colin Casimir, Ms Paula Clark, Dr Steve Hinshelwood, Dr Fay Katz, Dr Tracy Lester, Dr Ruth Lovering and Dr Carmel Teahan, and the Wellcome Trust and the Medical Research Council (UK) for their generous support.

References

Allen RC, Armitage RJ, Conley ME *et al.* (1993a) CD40 ligand gene defects responsible for X-linked hyper IgM syndrome. *Science,* **259,** 990–993.

Allen RC, Spriggs MK and Belmont JW. (1993b) Dinucleotide repeat polymorphism in the human CD40 ligand gene. *Hum. Mol. Genet.* **2,** 828.

Aruffo A, Farrington M, Hollenbaugh D *et al.* (1993) The CD40 ligand, gp39, is defective in activated T cells from patients with X-linked hyper-IgM syndrome. *Cell,* **72,** 291–300.

Ballabio A. (1993) The rise and fall of positional cloning? *Nature Genetics,* **3,** 277–279.

van Beusechem VW, Kukler A, Einerhand MPW, Bakx TA, van der Eb AJ, van Bekkum DW and Valerio D. (1990) Expression of human adenosine deaminase in mice transplanted with amphotropic retroviruses. *J. Exp. Med.* **172,** 729–736.

Bradley LD, Sweatman AK, Lovering RC, Jones AM, Morgan G, Levinsky RJ and Kinnon C. (1994) Mutation detection in the X-linked agammaglobulinemia gene, *BTK,* using single strand polymorphism analysis. *Hum. Mol. Genet.* **3,** 79–83.

Casimir CM and Teahan CG. (1994) The respiratory burst of neutrophils and its deficiency. In: *Immunopharmacology of Neutrophils* (eds PG Hellewell and TJ Williams). Academic Press, London, pp. 27–54.

Casimir CM, BuGhanim HN, Rodaway AR, Bentley DL, Rowe P and Segal AW. (1991) Autosomal recessive chronic granulomatous disease caused by deletion at a dinucleotide repeat. *Proc. Natl Acad. Sci. USA,* **88,** 2753–2757.

Collins FS. (1992) Positional cloning: let's not call it reverse anymore. *Nature Genetics,* **1,** 3–6.

DiSanto JP, Bonnefoy JY, Gauchat JF, Fischer A and de Saint Basile G.

(1993) CD40 ligand mutations in X-linked immunodeficiency with hyper IgM. *Nature,* **361,** 541–543.

Goodship J, Levinsky RJ and Malcolm S. (1989) Linkage of PGK1 to X-linked severe combined immunodeficiency (IMD4) allows predictive testing in families with no surviving male. *Hum. Genet.* **84,** 11–14.

Goodship J, Carter J, Espanol T, Boyd Y, Malcolm S and Levinsky RJ. (1991) Carrier detection in Wiskott–Aldrich syndrome: combined use of M27β for X-inactivation studies and as a linked probe. *Blood,* **77,** 2677–2681.

Graf D, Korthäuer U, Mages HW, Senger G and Kroczek RA. (1992) Cloning of TRAP, a ligand for DC40 on human T cells. *Eur. J. Immunol.* **22,** 3191–3194.

Grompe M. (1993) The rapid detection of unknown mutations in nucleic acids. *Nature Genetics,* **5,** 111–117.

Hendriks RW, Kraakman MEM, Craig IW, Espanol T and Schuurman RKB. (1990) Evidence that in X-linked immunodeficiency with hyper-immunoglobulinemia M the intrinsic immunoglobulin heavy chain class switch mechanism is intact. *Eur. J. Immunol.* **20,** 2603–2608.

Hollenbaugh D, Grosmaire LS, Kullas CD, Chalupny NJ, Braesch-Anderson S, Noelle RJ, Stamenkovic I, Ledbetter JA and Aruffo A. (1992) The human T cell antigen gp39, a member of the TNF gene family, is a ligand for the CD40 receptor: expression of a soluble form of gp39 with B cell co-stimulatory activity. *EMBO J.* **11,** 4313–4321.

Katz F, Lovering RC, Bradley LAD, Rigley KP, Brown D, Cotter F, Chessells JM, Levinsky RJ and Kinnon C. (1994) Expression of the X-linked agammaglobulinemia gene, *btk,* in B cell acute lymphoblastic leukemia. *Leukemia,* **8,** 574–577.

Kondo M, Takeshita T, Ishii N, Nakamura M, Watanabe S, Arai K-I and Sugamura K. (1993) Sharing of the interleukin-2 (IL-2) receptor γ chain between receptors for IL-2 and IL-4. *Science,* **262,** 1874–1877.

Korthäuer U, Graf D, Mages HW *et al.* (1993) Defective expression of T-cell CD40 ligand causes X-linked immunodeficiency with hyper-IgM. *Nature,* **361,** 539–541.

Lebkowski JS, Schain LR, Okrongly D, Levinsky RJ and Okarma TB. (1992) Rapid isolation of human CD34 hematopoietic stem cells: purging of human tumor cells. *Transplantation,* **53,** 1011–1019.

Lovering R, Middleton-Price HR, O'Reilly M-AJ *et al.* (1993) Genetic linkage analysis identifies new proximal and distal flanking markers for the X-linked agammaglobulinemia gene locus, refining its localization in Xq22. *Hum. Mol. Genet.* **2,** 139– 141.

Lovering RC, Sweatman A, Genet SA *et al.* (1994) Identification of deletions in the *btk* gene allows unambiguous assessment of carrier status in families with X-linked agammaglobulinemia. *J. Med. Genet.,* in press.

Markiewicz S, DiSanto JP, Chelly J *et al.* (1993) Fine mapping of the human SCIDX1 locus at Xq12–13.1 *Hum. Mol. Genet.* **2,** 651–654.

Miyajima A, Kitamura T, Harada N, Yokota T and Arai K-I. (1992) Cytokine receptors and signal transduction. *Annu. Rev. Immunol.* **10,** 295–331.

Noguchi M, Nakamura Y, Russell SM, Ziegler SF, Tsang M, Cao X and Leonard WJ. (1993a) Interleukin-2 receptor γ chain: a functional component of the interleukin-7 receptor. *Science,* **262,** 1877–1880.

Noguchi M, Yi H, Rosenblatt HM, Filipovich AH, Adelstein S, Modi WS, McBride OW and Leonard WJ. (1993b) Interleukin-2 receptor γ chain mutation results in X-linked severe combined immunodeficiency. *Cell,* **73,** 147–157.

Padayachee M, Feighery C, Finn A, McKeown C, Levinsky RJ, Kinnon C and Malcolm S. (1992) Mapping of the X-linked form of hyper IgM syndrome (HIGM1) to Xq26 by close linkage to HPRT. *Genomics,* **14,** 551–553.

Pawson T and Gish GD. (1992) SH2 and SH3 domains: from structure to function. *Cell,* **71,** 359–362.

Pelham A, O'Reilly M-AJ, Malcolm S, Levinsky RJ and Kinnon C. (1990) RFLP and deletion analysis for X-linked chronic granulomatous disease using the cDNA probe: potential for improved prenatal diagnosis and carrier determination. *Blood,* **76,** 820–824.

Porter CD, Parkar M, Levinsky RJ, Collins MKL and Kinnon C. (1993) X-linked chronic granulomatous disease: correction of NADPH oxidase defect by retrovirus-mediated expression of gp91-phox. *Blood,* **82,** 2196–2202.

Puck JM, Conley ME and Bailey LC. (1993a) Refinement of linkage of human severe combined immunodeficiency (SCIDX1) to polymorphic markers in Xq13. *Am. J. Hum. Genet.* **53,** 176–184.

Puck JM, Deschênes SM, Porter JC, Dutra AS, Brown CJ, Willard HF and Henthorn PS. (1993b) The interleukin-2 receptor γ chain maps to Xq13.1 and is mutated in X-linked severe combined immunodeficiency, SCIDX1. *Hum. Mol. Genet.* **2,** 1099–1104.

Ramesh N, Fuleihan R, Ramesh V, Lederman S, Yellin MJ, Sharma S, Chess L, Rosen FS and Geha RS. (1993) Deletions in the ligand for CD40 in X linked immunoglobulin deficiency with normal or elevated IgM (HIGMX-1). *Int. Immunol.* **5,** 769–773.

Rawlings DJ, Saffran, DC, Tsukada S *et al.* (1993) Mutation of unique region of Bruton's tyrosine kinase in immunodeficient XID mice. *Science,* **261,** 358–361.

Ren R, Mayer BJ, Cicchetti P and Baltimore D. (1993) Identification of a ten-amino acid proline rich SH3 binding site. *Science,* **259,** 1157–1161.

Royer-Pokora B, Kunkel LM, Monaco AP, Goff SC, Newburger PE, Baehner RL, Cole FS, Curnutte JT and Orkin SH. (1986) Cloning the gene for an inherited human disorder – chronic granulomatous disease – on the basis of its chromosomal location. *Nature,* **322,** 32–38.

Russell SM, Keegan AD, Harada N *et al.* (1993) Interleukin-2 receptor γ chain: a functional component of the interleukin-4 receptor. *Science,* **262,** 1880–1883.

Spriggs MK, Armitage RG, Strockbine L, Clifford KN, Macduff BM, Sato TA, Maliszewski CR and Fanslow WC. (1992) Recombinant human CD40 ligand stimulates B cell proliferation and immunoglobulin E secretion. *J. Exp. Med.* **176,** 1534–1550.

Taniguchi T and Minami Y. (1993) The IL-2/IL-2 receptor system: a current overview. *Cell,* **73,** 5–8.

Thrasher A, Chetty M, Casimir C and Segal AW. (1992) Restoration of superoxide generation to a chronic granulomatous disease-derived B-cell line by retrovirus mediated gene transfer. *Blood,* **80,** 1125–1129.

Thomas JD, Sideras P, Smith CIE, Vorechovsky I, Chapman V and Paul WE. (1993) Colocalization of X-linked agammaglobulinemia and X-linked immunodeficiency genes. *Science,* **261,** 355–358.

Tsukada S, Saffron DC, Rawlings DJ *et al.* (1993) Deficient expression of a B cell cytoplasmic tyrosine kinase in a human X-linked agammaglobulinemia. *Cell,* **72,** 279–290.

Vetrie D, Vorechovsky I, Sideras P *et al.* (1993) The gene involved in X-linked agammaglobulinaemia is a member of the *src* family of protein-tyrosine kinases. *Nature,* **361,** 226–233.

Waksman G, Shoelson SE, Pant N, Cowburn D and Kuriyan J. (1993) Binding of a high affinity phosphotyrosyl peptide to the *src* SH2 domain: crystal structures of the complexed and peptide-free forms. *Cell,* **72,** 779–790.

de Weers M, Mensink RGJ, Kraakman MEM, Schuurman RKB and Hendriks RW. (1994) Mutation analysis of the Bruton's tyrosine kinase gene in X-linked agammaglobulinemia: identification of a mutation which affects the same codon as is altered in immunodeficient *xid* mice. *Hum. Mol. Genet.* **3,** 161–166.

4

Duchenne muscular dystrophy

S.C. Brown and G. Dickson
Department of Experimental Pathology, UMDS Guys Hospital, London, UK

1. Pathology and genetics of Duchenne muscular dystrophy

Duchenne muscular dystrophy (DMD) and its milder form, Becker muscular dystrophy (BMD), are allelic X-linked muscle-wasting disorders in man. DMD affects 1/3500 males, one-third of which are sporadic cases with no previous family history. In DMD, the disease is progressive, with proximal muscle weakness leading to loss of ambulation at around 11 years of age. The regeneration of damaged fibres eventually fails to compensate for the recurrent phases of degeneration, and is further hampered by the gradual deposition of fibrotic connective tissue at severely affected sites. Death due to respiratory failure usually occurs by the third decade (Emery, 1993). At present there is no effective treatment to prevent or limit the progressive skeletal and cardiac weakness that ultimately leads to severe disability in DMD. The phenotype characteristic of BMD is, however, much more varied, with some patients never losing the ability to walk. Both DMD and BMD are now known to be caused by mutations in the gene that codes for dystrophin (Ahn and Kunkel, 1993; Hoffman *et al.*, 1987).

The DMD/BMD gene was among the first to be characterized by positional cloning, without any prior knowledge of the biochemical defect. The genomic locus extends over 2400 kb, of which approximately 14 kb is transcribed into an mRNA comprising 79 exons (Roberts *et al.*, 1993). An approximately 11 kb open reading frame (ORF) is translated into a protein which, in striated muscle, is composed of 3685 amino acids and has a molecular weight of 427 kDa (Koenig *et al.*, 1988). More recently, however, a range of alternative splicing events, and transcriptional initiation from at least five distinct promoter elements, have been described (Ahn and Kunkel, 1993). The DMD gene is thus a very complex transcriptional unit, giving rise to a wide diversity of isoform structure and expression in a range of tissues (Barnea *et al.*, 1990; Blake *et al.*, 1993; Feener *et al.*, 1989; *Figure 1*).

Approximately 70% of cases of DMD, and over 80% of cases of BMD,

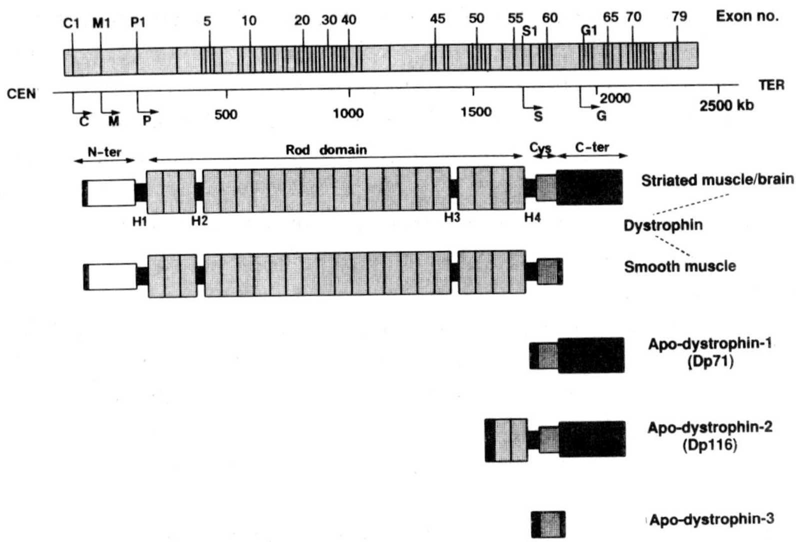

Figure 1. Diagrammatic representation of the DMD gene and its products. The 79 exons span 2.4 million bases. At least five distinct promoters drive the cell type-specific expression of different dystrophin isoforms. The C (cortical), M (muscle) and P (Purkinje cell) promoters all express full-length forms of dystrophin, whilst the S (Schwann cell) and G (general or glial) dystrophin promoters encode the C-terminal apo-dystrophins Dp116 and Dp71, respectively. The diagram of the domain organization of dystrophin shows: (1) the N-terminal region (N-ter) which binds to F-actin; (2) the long internal region consisting of 24 homologous repeats of approximately 109 amino acids which form a long rod-like domain; (3) the cysteine-rich region (Cys) which is homologous to the Ca^{2+}-binding EF hand region of α-actinin, and (4) the C-terminal domain (C-ter) which is homologous only to the dystrophin-related protein, utrophin. The four putative hinge structures in the rod domain (H1–H4), and the occurrence of alternatively spliced regions at the extreme N- and C-termini of dystrophin and apo-dystrophins are indicated as shaded areas.

appear to be due to gene deletions. Deletions can occur almost anywhere within the dystrophin gene. However, most seem to locate to two 'hot-spots', either around exons 45–55 (1200 kb from the 5′ end of the gene) or close to the first 20 exons (500 kb from the 5′ end of the gene). Deletion size has no influence over whether the phenotype is Duchenne or Becker. Instead, in the majority of cases, this depends on the influence of partial gene deletions on the ORF of triplet codons in the mRNA transcript (Koenig *et al.*, 1989; Monaco *et al.*, 1988). Deletions causing a shift in the ORF (frameshift deletions) result in the Duchenne phenotype, presumably due either to metabolic instability of the mutant mRNAs or proteins or, most likely, to functional inactivity of the latter (Hoffman and Kunkel, 1989). By contrast, those which maintain the ORF (in-frame deletions) give rise to milder Becker-type disease, associated with expression of mutant forms of dystrophin which, although internally deleted and of a lower molecular weight, are proposed to exhibit variable but significant functional activity. Exceptions to this so-called 'reading frame' hypothesis are thought to be due to variant alternative splicing patterns or skipping of juxtaposed exons, which effectively overcome the effect on the ORF of the original deletion mutation.

Gene duplications account for 5–10% of DMD cases, and are probably the result of unequal sister chromatid exchange (Emery, 1993). Point mutations causing a stop codon or amino acid substitution can also lead to severe disease. It is possible that mutations involving amino acid substitutions perturb subunit interactions between dystrophin molecules or between dystrophin and other proteins, and their identification should provide insights critical to the understanding of dystrophin structure and function (Roberts *et al.*, 1992).

In addition to the human disease, several animal models of dystrophin deficiency have been described, namely the *mdx* mouse (Bulfield *et al.*, 1984), *xmd* dog (Valentine *et al.*, 1990) and the dystrophic cat (Gaschen *et al.*, 1992). The consequences of dystrophin deficiency in each of these species is very different, with the *mdx* mouse showing symptoms that are milder than those observed in patients with DMD. However, many skeletal muscles and, in particular, the diaphragm and intercostal muscles of older *mdx* mice, do eventually exhibit a progressively severe and fibrotic myopathy similar to the human disease (Stedman *et al.*, 1991). In the *xmd* dog, the myopathy follows a similar course to DMD in man, with progressive muscular weakness, hypertrophy of defined muscle groups and premature death due to respiratory insufficiency (Valentine *et al.*, 1990). The cat model shows marked diaphragmatic and glossal hypertrophy but, unlike the human disease and the *xmd* dog, there is no progressive loss of muscle fibres, fibrosis or weakness, despite the deficiency of dystrophin.

2. Structure and function of dystrophin

Amino acid sequence analysis indicates that dystrophin is comprised of four contiguous domains (Koenig *et al.*, 1988). These are: (1) an N-terminal region (240 amino acids) which is similar to the conserved actin-binding domain of α-actinin, spectrin and *Dictyostelium* actin-binding protein 120, and which binds to filamentous actin (Hemmings *et al.*, 1992); (2) a large domain comprising 24 triple-helical sequences, which resemble the repeats in spectrin and may be elastic; (3) a cysteine-rich region between residues 3080 and 3360 which shows significant homologies to the C-terminal domain of α-actinin in *Dictyostelium discoideum* and may contain two calcium-binding sites, although this is by no means proven, and; (4) a sequence of 420 amino acids at the C-terminal end which appears to be unique apart from a similarity to the C-terminus of the dystrophin-related protein (DRP) utrophin encoded on chromosome 6 (see below). Thus, it has been suggested that dystrophin belongs to the spectrin superfamily of proteins (Ahn and Kunkel, 1993).

As shown in *Figure* 2, dystrophin locates to the cytoplasmic face of the normal adult muscle fibre membrane (Byers *et al.*, 1991), where it comprises approximately 5% of the membrane cytoskeleton (Ohlendieck and Campbell, 1991). In addition to being present at both the neuromuscular and myotendinous junctions (Byers *et al.*, 1991; Ohlendieck *et al.*, 1991), dystrophin is specifically enriched in three distinct domains along the sarcolemma, namely in transverse rib-like structures overlying both the I bands and M lines of peripheral sarcomeres, and in occasional strands running along the longitudinal axis of the fibre. This distribution is similar to the so-called membrane costamere lattice in skeletal myofibres which is thought to mediate the transmembrane interaction of contractile elements with the extracellular basal lamina (Dickson *et al.*, 1992; Porter *et al.*, 1992).

Biochemically, dystrophin has indeed been clearly shown to associate with a membrane-spanning oligomeric complex of six sarcolemmal proteins, four of which are glycoproteins (Ervasti and Campbell, 1991; Ohlendieck *et al.*, 1991; *Figure* 3). This so-called dystrophin-associated glycoprotein complex (DAGC) appears to interact with the cysteine-rich and C-terminal domains of dystrophin (Ervasti and Campbell, 1991;

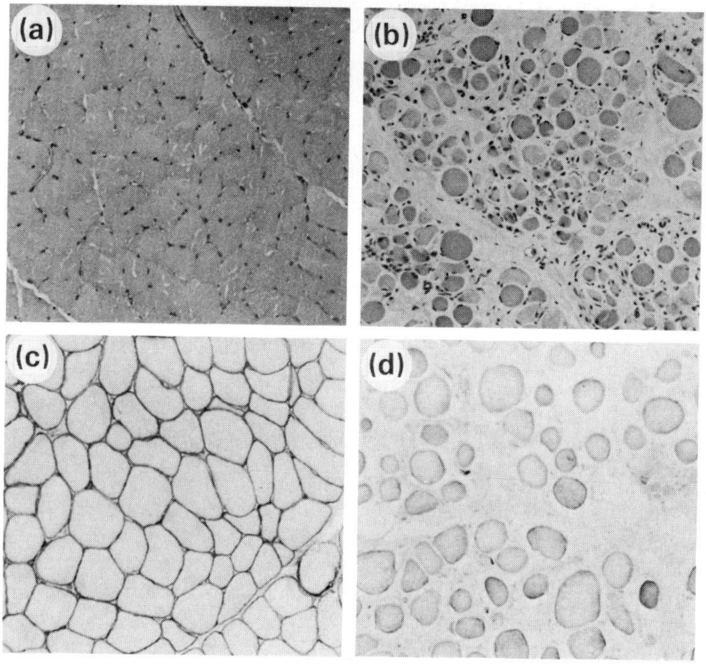

Figure 2. A comparison of haematoxylin- and eosin-stained transverse sections of skeletal muscle from a control (a) and DMD patient (b), which clearly show the extensive fibre loss and fibrosis characteristic of DMD. The immunocytochemical stains in the lower panels show the location of dystrophin at the muscle fibre sarcolemma in normal skeletal muscle (c), and its virtual absence in DMD muscle (d).

Suzuki *et al.*, 1992), whilst the N-terminus of dystrophin binds to cortical actin microfilaments. Furthermore, the 156 kDa component of the DAGC is, in fact, a peripheral surface proteoglycan (Ervasti and Campbell 1991) which binds to laminin in the extracellular matrix (Dickson *et al.*, 1992; Ibraghimov-Beskrovnaya *et al.*, 1992), thereby potentially linking the actin-based sarcolemmal cytoskeleton with the extracellular matrix.

The absence of dystrophin in the muscle of *mdx* mice and DMD patients also results in a marked reduction of the DAGC, and autosomally inherited muscular dystrophy in humans and in hamsters has been shown to arise due to mutations in at least one of its components (50 kDa; Matsumura *et al.*, 1992b; Robards *et al.*, 1993). The loss of linkage between the actin cytoskeleton and the extracellular matrix may act to compromise sarcolemmal integrity, so as to render the muscle

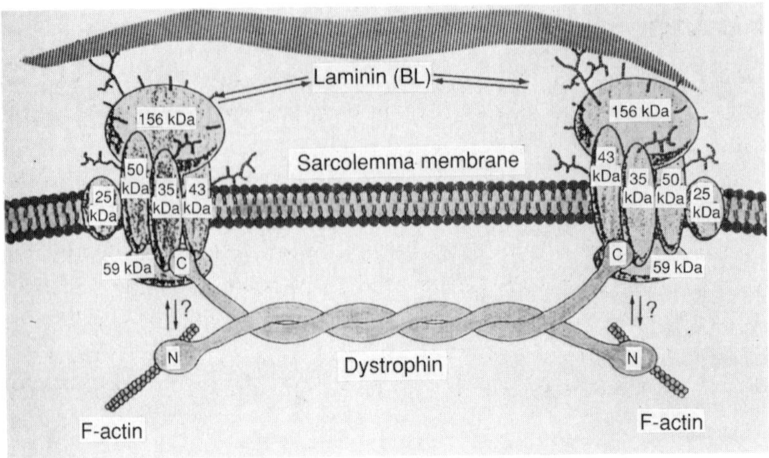

Figure 3. A model proposed by Ervasti and Campbell (1991) for the interaction of dystrophin with the plasma membrane, in which the N-terminal region interacts with cytoskeletal actin filaments while the cysteine-rich and C-terminal domains bind to a complex of membrane-spanning glycoproteins (DAGC). The peripheral dystroglycan molecule (156 kDa component of the DAGC) is thought to provide a link with laminin in the myofibre basal lamina (BL). Reproduced from Ervasti and Campbell (1991) with permission from Cell Press.

fibre more susceptible to mechanical damage during normal contractile activity (Menke and Jockusch, 1991; Stedman *et al.*, 1991; Weller *et al.*, 1990). Alternatively, dystrophin, the DAGC and other, as yet undefined, factors may act to maintain some vital spacial or regulatory function of other crucial myofibre membrane proteins, such as those involved in calcium homeostasis (Turner *et al.*, 1991). This latter possibility is supported by the observation of elevated levels of free calcium measured in myotubes derived from the muscle of *mdx* mice and DMD patients, compared with normal myotubes (Bakker *et al.*, 1993; Franco and Lansman, 1990; Turner *et al.*, 1991). However, to date, evidence defining the precise molecular, or indeed cellular, function of dystrophin, and the mechanism underlying the fibre necrosis which occurs in its absence, remains equivocal and, at best, circumstantial. Molecular complementation of dystrophin deficiency by transfer of recombinant genes, mRNA or protein itself, or acute knock-out of dystrophin activity by microinjection of anti-sense DNA, blocking antibodies or peptide reagents, offer the best experimental paradigms to address these functional questions.

3. Gene therapeutic targets in DMD

The cloning and characterization of the DMD gene, its cDNA and recognition of the structural and developmental relationship between dystrophin and utrophin, have led to proposals for gene activational or gene transfer strategies as possible therapeutic approaches to treat the main symptoms of the disease (*Figure 4*).

As already indicated, dystrophin or closely related isoforms are present in both striated and smooth muscle cells, and in CNS neuronal populations, and there is widespread expression of apo-dystrophins in a range of tissues (Ahn and Kunkel, 1993; Blake *et al.*, 1993; Hoffman *et al.*, 1987). However, the major clinical pathophysiology in DMD arises,

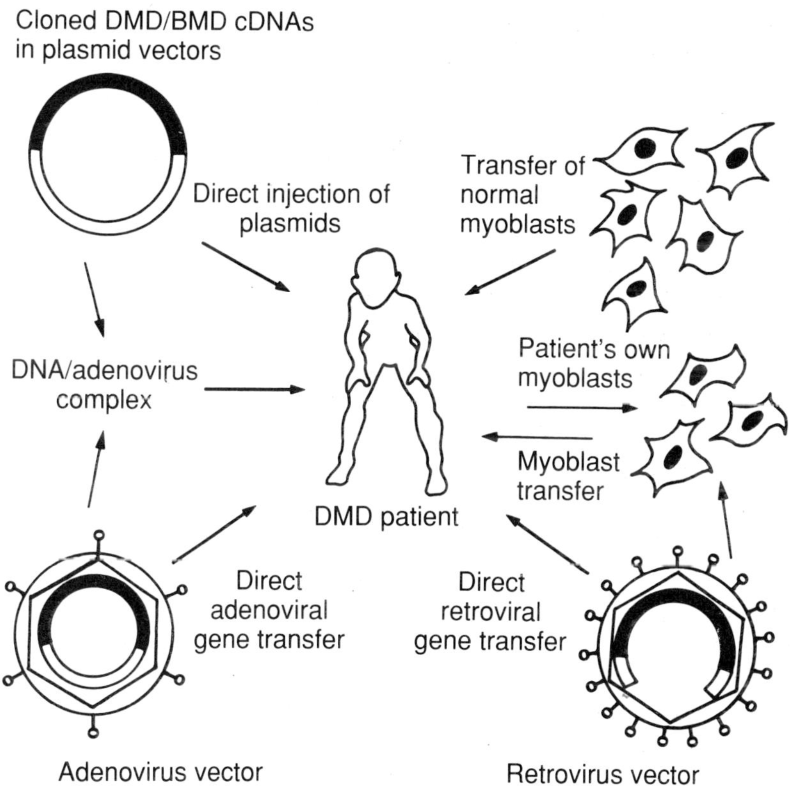

Figure 4. Diagram showing the options for gene therapy in DMD which include myoblast transfer, direct and indirect retroviral and adenoviral gene transfer, or the direct injection of plasmids containing cloned DMD/BMD cDNAs.

in the main, from the progressive degeneration and loss of dystrophin-deficient skeletal myofibres, which becomes life-threatening as respiratory muscle function is compromised. The development of cardiomyopathic features in later stages of the disease is also a major cause of mortality (Emery, 1993; Wessel, 1990). Thus, dystrophin gene transfer into selected skeletal, and possibly cardiac, muscles could be considered a minimum therapeutic target in designing gene delivery systems of clinical utility.

Several characteristics of skeletal muscle are important when considering it as a target for gene therapy. Muscle forms developmentally from the fusion of mononucleated myoblasts to produce myofibres which, in the adult, contain many thousands of nuclei. All fibre nuclei are post-mitotic, so that any gene transfer directly into the nuclei of the fibre will not subsequently be 'diluted out' by cell division. However, the nuclear domain (i.e. that area of the fibre which contains the protein products of a single nucleus) tends to be restricted for structural and membrane proteins (Pavlath *et al.*, 1989), including dystrophin (Noursadeghi *et al.*, 1993), which implies that a significant proportion of fibre nuclei may require a copy of the dystrophin gene to improve the functional capacity of the muscle fibre as a whole. In this respect, dystrophin expression in so-called manifesting heterozygote female carriers of DMD can be as much as 30% of normal levels (Sewry *et al.*, 1993). However, in other situations where, for example, promoter mutations severely reduce dystrophin expression in skeletal muscle but not the integrity or general distribution of the molecule itself, a relatively mild skeletal muscle phenotype is observed (Boyce *et al.*, 1991).

In addition to direct gene transfer to terminally differentiated skeletal myofibres, genes could be introduced via the mononucleated stem or progenitor cell populations of skeletal muscle, since proliferation and fusion of these so-called satellite cells with myofibres occurs during the normal growth and repair of muscle. Satellite cells lie between the sarcolemma and basal lamina of myofibres, and are mitotically active in muscle which is either fast-growing, or undergoing repair. Since fibre regeneration is one of the earliest features of DMD (Emery, 1993), satellite cells carrying a biologically active recombinant dystrophin gene, introduced either directly *in vivo* or via cultured cells in an *ex vivo* approach, could be a very effective way of introducing dystrophin into necrotic areas of the muscle. However, since these cells may undergo many rounds of cell division before eventually being incorporated into the muscle fibre (Campion, 1984), any gene transferred would have to be stably integrated into the genome. Nonetheless, since between 50 and 99% of the total nuclear content of the muscle accumulates post-natally, this population of cells could provide an important long-term repository for transgene expression. As discussed in Section 5, one approach to effective gene transfer which has been examined in clinical trials is the grafting of normal healthy satellite cells from an appropriate donor to repopulate diseased, genetically deficient muscle in a classic cell transplantation procedure. This similarity in principle between somatic gene transfer and cell transplantation underscores the close relationship that exists between the ethical bases of the two procedures.

4. Recombinant dystrophin genes

The DMD gene forms a complex transcriptional unit containing specific neural and striated muscle promoter structures (Barnea *et al.*, 1990). The large size of the native gene structure precludes direct cloning within conventional plasmid, phage or eukaryotic viral vectors. However, the entire gene has now been incorporated within a yeast artificial chromosome (YAC) construct (Den Dunnen *et al.*, 1992), and the microinjection, liposome- or protoplast-mediated introduction of YAC constructs into eukaryotic cells has been shown to be a feasible, although highly inefficient, route of gene transfer (Gnirke and Huxley, 1991; Strauss and Jaenisch, 1992). These observations suggest that, in the future, it may be possible to manipulate and transfer the entire, or large regulatory segments of the, normal DMD locus into dystrophin-deficient cells. However, since introducing a significant number of such genes into muscle fibres may be too difficult at the present time, most effort has been invested in constructing and expressing recombinant dystrophin genes based on cDNA clones.

Full-length recombinant cDNAs have been described encoding human (Dickson *et al.*, 1991; *Figure 5*) and mouse (Lee *et al.*, 1991) skeletal muscle dystrophin. For the human cDNA, the starting point was a panel of five overlapping partial cDNA clones which were pieced together by a sub-cloning strategy to yield a final 12 kb product spanning the entire ORF of the human mRNA as it is expressed in adult human muscle. For the mouse cDNA, specific reverse transcription primers, based on the sequence of mouse dystrophin mRNA, were used to produce directly two large overlapping cDNA segments corresponding to 5′ and 3′ halves of the mRNA, which were then joined centrally. Both the cDNAs were initially cloned into constitutive high-level

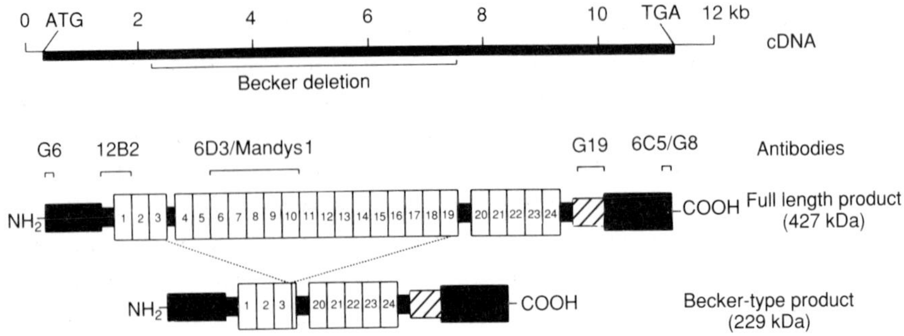

Figure 5. Diagrammatic representation of dystrophin cDNA and polypeptides showing the position of a large 5.1 kb central in-frame deletion found to be associated with a mild BMD phenotype. The domain map shows the extent of the deletion on the central rod domain of which over 63% is deleted.

eukaryotic expression plasmids and, following transfection into 3T3 fibroblasts and COS cells, were shown to encode recombinant dystrophin polypeptides of appropriate size (427 kDa) corresponding to the major form of dystrophin observed in adult muscle. Furthermore, full-length mouse (Cox *et al.*, 1993) and human cDNAs (Dickson *et al.*, unpublished data) have now been expressed in transgenic *mdx* mice using muscle-specific M-creatine kinase or skeletal actin promoters. Both the human and mouse recombinant dystrophins correct the early necrotic and later fibrotic pathophysiology in *mdx* skeletal muscle.

While active genes encoding functional wild-type dystrophin are available, they present some problems for somatic gene transfer studies in that their size (>12 kb) is clearly above the packaging limits for adenovirus-based vectors and is widely believed, but not yet proven, to be too large for efficient retroviral vector encapsidation. BMD patients produce forms of dystrophin which, although of a lower molecular weight, are at least partially functional. One such patient has been shown to have a deletion of 5106 nucleotides of coding sequence, that removes 63% of the rod domain, but which still results in a relatively mild phenotype (England *et al.*, 1990). This particular patient is still alive and ambulatory at the age of 69 years at the time of writing (K.E. Davies, personal communication). As we shall consider in detail later, recombinant cDNA gene constructions based on this patient's genotype thus offer the potential to express a highly functional mini-dystrophin molecule within the context of either an adenovirus- or retrovirus-based gene transfer vector. However, to be of use in gene therapy, this BMD-based minigene had to be shown to encode a functional product. The cDNA corresponding to this patient has therefore been reconstructed (Acsadi *et al.*, 1991) and examined *in vitro* by plasmid-, retroviral- and adenoviral-mediated somatic gene transfer into *mdx* muscle (Acsadi *et al.*, 1991; Dunckley *et al.*, 1992, 1993; Ragot *et al.*, 1993; Vincent *et al.*, 1993) and by microinjection into fertilized mouse ova to produce transgenic *mdx* mice (Wells *et al.*, 1992). In each case, the truncated dystrophin molecule localized to the plasma membrane and, in the transgenic mice, its expression was correlated with a significant reduction in muscle fibre necrosis and regeneration, and prevented the later severe fibrotic changes that normally occur in *mdx* diaphragm (Wells *et al.*, 1992; Dickson *et al.*, unpublished data; *Figure 6*).

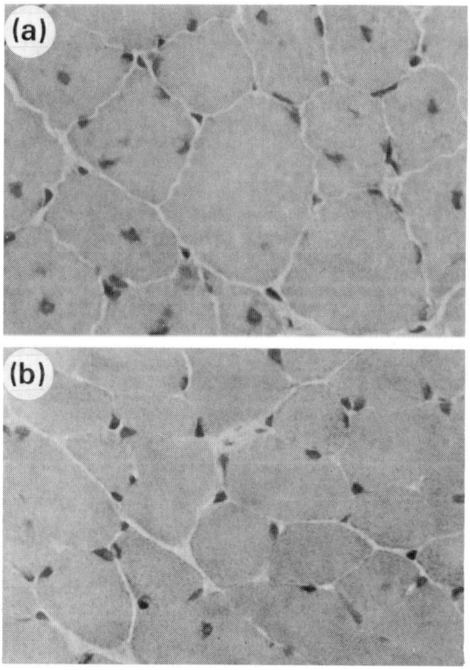

Figure 6. Haematoxylin- and eosin-stained transverse sections of muscle taken from 6-week-old control *mdx* mice (a) and transgenic *mdx* mice expressing recombinant human dystrophin (b). The centrally located nuclei characteristic of regenerated *mdx* muscle (a) contrasts with the transgenic mouse, where most of the nuclei are peripherally located (b), as in the normal animal.

5. Myoblast transplantation

As already considered, skeletal muscle cells are multinucleated syncytia, and one mechanism for the introduction of novel genetic information into this tissue is by myoblast transplantation. This approach involves the introduction of normal donor or genetically engineered myoblast satellite cells into the diseased muscle to correct the genetic deficiency locally. Theoretically, the injection of these cells should lead to the formation of mosaic fibres (i.e. fibres in which at least some of the nuclei produce dystrophin). The source of the transplanted cells could be the muscles of the patients themselves (autologous myoblast transfer), in which case satellite cells would be isolated from the muscle, transduced with the recombinant dystrophin gene and then re-introduced. Alternatively, normal myoblasts could be obtained from an unrelated donor (heterologous myoblast transfer).

Autologous myoblast transfer has the advantage that it avoids any immunological rejection associated with the cells themselves (although not with dystrophin itself, if it proved to be antigenic). However, satellite cells derived from DMD muscle may have already exhausted their proliferative potential, due to the repetitive cycles of degeneration and regeneration that have already occurred within the diseased muscle. There would consequently be a problem in obtaining a sufficient number of cells to inject, especially if a number of injections were envisaged.

Myoblast transfer in human subjects has, to date, yielded poor or, at best, extremely contentious results. In one clinical experiment, dystrophin mRNA was detectable by nested polymerase chain reaction (PCR) amplification several weeks following myoblast implantation and, in another, improved force generation was reported. The significance of both these studies and, in the latter case, the validity of the data itself, has been the subject of intense dispute. However, in the best controlled double-blind study to date, no correction of dystrophin deficiency was observed, as judged either by improved force generation and muscle dystrophin content, or the presence of donor genomic DNA and dystrophin mRNA (Engel, 1993; Karpati *et al.*, 1993b). The low efficiency of myoblast transfer has been attributed to a number of factors, including the poor mobility of the injected cells through connective tissue barriers in the muscle, failure of myoblasts to fuse with non-regenerating muscle fibres and a possible impairment of the proliferative capacity of the donor myoblasts (Karpati *et al.*, 1993b). In addition, in a recent study where paternal myoblasts were transplanted, it was shown that there was an increase in HLA class 1 antigen expression on the muscle fibres of the injected muscle (Karpati *et al.*, 1993). Therefore, it is possible that, despite immunosuppressant therapy, some donor cells and hybrid muscle fibres expressing HLA class 1 alloantigens were eliminated within the first few hours or days by alloreactive cytotoxic T cells and natural killer cells (Engel, 1993). These problems, together with the general inaccessibility of some muscles, such as the diaphragm and heart, suggest that the clinical usefulness of myoblast transplantation is limited.

6. Activated expression of 'embryonic' dystrophin: the utrophin molecule

One variant to classical gene transfer therapy for inherited recessive disease has been the proposal that reactivation of an embryonic or fetal isoform might complement the deficiency of a mutated adult gene, as has been long postulated for the β-globin gene. Utrophin, or DRP, is the product of a large gene spanning almost 1 Mb on the long arm of the human chomosome 6 and mouse chromosome 10 and is thought, at least in skeletal muscle, to be an embryonic isoform of dystrophin (Love *et al.*, 1993). Utrophin shows extensive homology with dystrophin, especially at the N-terminus and the cysteine-rich and C-terminal domains, suggesting that it may bind to both actin and the DAGC complex (Blake *et al.*, 1993). Utrophin is, in fact, expressed widely in many adult tissues, but is concentrated at the post-synaptic membrane of the neuromuscular junction, the myotendinous junction, peripheral nerves and vasculature of adult skeletal muscle tissue. However, it is not normally expressed at the extra-junctional sarcolemma of adult skeletal muscle (Blake *et al.*, 1993; Love *et al.*, 1993), but general sarcolemmal staining has been observed in the regenerating muscle fibres of DMD/BMD patients and *mdx* mice (Karpati *et al.*, 1993a; Love *et al.*, 1993).

In skeletal muscle, utrophin gene expression is clearly developmentally regulated, since the levels of its mRNA are high in human and mouse fetal tissue but low in the adult (Walsh *et al.*, 1989). This is reflected in normal fetal muscle, where utrophin shows a maximum level of expression at 17–18 weeks gestation, which subsequently decreases as the expression of dystrophin increases (Clerk *et al.*, 1993). As a consequence of this pattern of expression, together with the extensive sequence homology between utrophin and dystrophin, there has been much interest as to whether utrophin could compensate for the absence of dystrophin. For this to be possible, utrophin would need to be functionally very similar to dystrophin. This may indeed be the case, as utrophin appears to bind to the DAG complex, or at least to one that is antigenically similar (Matsumura *et al.*, 1992a). In addition, there seems to be a tendency for those *mdx* muscles showing little pathology to produce high levels of utrophin compared to those which are more severely affected by the myopathy (Matsumura *et al.*, 1992a). However, it remains to be shown whether utrophin, if expressed in adult dystrophin-deficient myofibres, can effectively prevent or reduce muscle damage or, indeed, if any pharmacological mechanism for activation of the utrophin gene in mature muscle can be identified.

7. Physical gene transfer strategies

A range of procedures have been utilized to induce mammalian cells to take up and express functional recombinant genes. These have included physical DNA-mediated transfection, involving the direct intramuscular administration of pure plasmid DNA (Acsadi *et al.*, 1991; Wolff *et al.*, 1990), and the introduction of DNA complexed with surface-receptor ligands, adenoviral particles, cationic liposomes or ballistic microprojectiles (Dickson and Dunckley, 1993; Dunckley and Dickson, 1994; Mulligan, 1993; Zhu *et al.*, 1993). Gene transfer has also been achieved by exposure of cells to rapid pulses of high voltage current to induce transient membrane permeability (electroporation), and microinjection procedures (Graessmann and Graessmann, 1983). The efficiency of physical transfection can exceed 1% in suitable recipient cells *in vitro*, and the direct transfer of genes into whole animals *in vivo* has also been shown.

The direct transfer of dystrophin cDNA-based gene constructs has been achieved in both fibroblast and myoblast cell lines. The recombinant dystrophin has been shown to be of an appropriate size, and appears to localize to the cell surface (Dickson *et al.*, 1991). Similar results are obtained after transfection of primary myoblast cultures derived from the dystrophin-deficient mouse (*mdx*) and human (DMD) patients, where the transgene product locates to plasma membrane of differentiated myotubes (Dickson *et al.*, unpublished data). However, it is the ability of recombinant dystrophin to associate with the skeletal muscle sarcolemma of *mdx* mice, following the intramuscular injection of DNA constructs, that provides the most convincing evidence that recombinant dystrophin is able to localize correctly within the muscle fibre (Acsadi *et al.*, 1991; *Figure 7*). More importantly, the apparent reduction in myofibre regeneration compared with control dystrophin-negative myofibres (Acsadi *et al.*, 1991) strongly suggests that the recombinant dystrophin protein is biologically active.

The feasibility of using direct injection as a means of introducing recombinant DNA into skeletal muscle arose from the observations made by Wolff *et al.* (1990, 1991), who noted that the injection of purified plasmid containing the *Escherichia coli* β-galactosidase or firefly luciferase genes into rodent muscle resulted in relatively long-term gene expression in individual myofibres (Wolff *et al.*, 1990, 1991). The cellular and molecular basis of this form of gene transfer remains poorly understood. However, in the case of the luciferase gene, injected

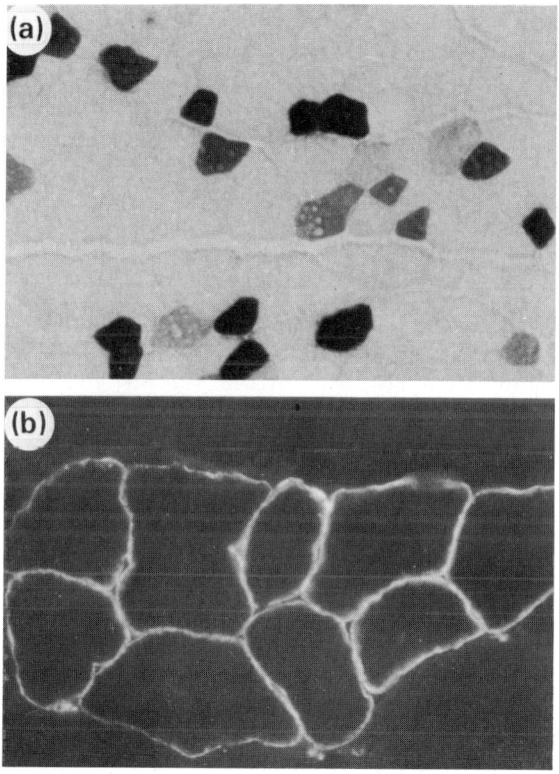

Figure 7. Transverse sections of *mdx* mouse muscle following direct injection of purified plasmid gene constructs expressing either the *E. coli* β-galactosidase gene (X-gal staining, a) or human mini-dystrophin cDNA (dystrophin immunostaining, b) which are under transcriptional control of Rous sarcoma virus (RSV) and Moloney leukaemia virus (MoMLV) promoters, respectively.

DNA appears to persist in skeletal muscle, at least, in an unintegrated extra-chromosomal state, and expression continues at relatively constant levels for at least 12 months. Similar experiments with dystrophin gene constructs injected into *mdx* skeletal muscle have resulted in recombinant gene products localized at the sarcolemma of myofibres in clusters around the injection site for at least up to 3 months (Acsadi *et al.*, 1991; Dickson and Dunckley, 1993; J. Wolff, personal communication).

Unfortunately, direct plasmid injection procedures for somatic gene transfer into skeletal muscle have proved to be of too low efficiency for potential clinical use in DMD, since experiments with the *mdx* mouse show that only 1–2% of myofibres display a dystrophin-positive phenotype following a single injection of DNA into the rectus femoris (Acsadi *et al.*, 1991; Dickson and Dunckley, 1993). However, this phenomenon has gained considerable credence as an approach towards DNA-mediated genetic immunization, or the ectopic expression of circulating recombinant products. One of the factors influencing the efficiency of this technique is the developmental and physiological status of targeted skeletal muscle, which may be amenable to manipulation (Davis *et al.*, 1993; Wells, 1993; Wells and Goldspink, 1992). In addition, *in vivo* DNA delivery may be enhanced by the use of fusogenic liposomes (Zhu *et al.*, 1993) or biolistic microprojectiles as DNA carriers (Williams *et al.*, 1991). It seems likely that unique advantages of simplicity and safety, and the unusual stability of direct DNA-mediated gene transfer into skeletal muscle, will lead to the development of increasingly efficient procedures.

8. Virus-mediated gene transfer strategies

In view of the low level of efficiency of direct DNA-mediated gene transfer, various viral systems have been adapted to develop potentially high efficiency somatic gene transfer vectors. These include vectors based on herpes simplex virus, vaccinia and adeno-associated virus (Mulligan, 1993) and, more extensively, replication-defective adenoviral (Berkner, 1992) and retroviral systems (Eglitis *et al.*, 1985). With the exception of the retroviral systems, these vectors have been designed and utilized for transient, but high efficiency, gene transfer studies (Miller, 1992; Mulligan, 1993).

The biology of the retroviral life-cycle displays many features which potentially can be used to enhance the efficiency and stability of somatic gene transfer (Eglitis *et al.*, 1985; Miller, 1992). Replication-defective, yet infectious, recombinant retroviral particles are able to infect a wide range of cell types from a diverse number of species, including humans (*Figure 8*). Following infection, the RNA within the viral genome is reverse transcribed to DNA, and permanently integrated into the host cells' chromosomes, from where it can be expressed at high efficiency. Efficient integration, however, requires host cell division and retroviral vectors can target genes to skeletal muscle only through activated stem cells in sites of induced or pathological regeneration. In general, vectors based on murine leukaemia virus have been used in human gene therapy trials, although alternative hybrid systems with vesicular stomatitis virus and those offering increased viral particle stability have been described recently (Burns *et al.*, 1993), as have HIV-based vectors which theoretically will also transduce DNA into quiescent non-dividing cells (Poznansky *et al.*, 1991).

There has also been considerable recent interest in the use of adenovirus vectors based on serotypes 2 and 5 for direct *in vivo* gene transfer into lung, liver, heart and skeletal muscle (Barr *et al.*, 1994; Quantin *et al.*, 1992; Rosenfeld *et al.*, 1992; Stratford-Perricaudet *et al.*, 1992; Zabner *et al.*, 1993; *Figure 9*). In this situation, the transferred adenoviral genome appears to remain as an episomal (i.e. non-integrated) element in successfully infected cells until lost in the course of normal cell turnover or, more likely, as the result of host immune responses. Approximately 80% of the human population are seropositive to these relatively benign pathogens. When administered to neonatal mice, human adenoviral vectors can apparently mediate long-term expression of transgenes, presumably due to induced immune tolerance of the mice to viral antigens. In adult rodents, inflammatory responses and the emergence of neutralizing antibodies have been reported (Smith *et al.*, 1993). However, if problems associated with host immune surveillance can be circumvented, very high titre recombinant adenovirus preparations can be used to transduce skeletal muscle myofibres at high efficiency *in vivo*, with no requirement for dividing cells.

One common limiting feature of currently available retroviral and adenoviral vector systems is that both have a limited ability to accept foreign DNA. At the present time, the amount of foreign DNA which can be stably incorporated in an adenoviral vector is less than 7 kb. The packaging constraint on MLV-based retroviral systems remains ill-defined but, if comparable to feline leukaemia virus, is liable to be <12 kb, with a minimum of about 2 kb required for essential packaging, integration and transcription elements (Eglitis *et al.*, 1985). This

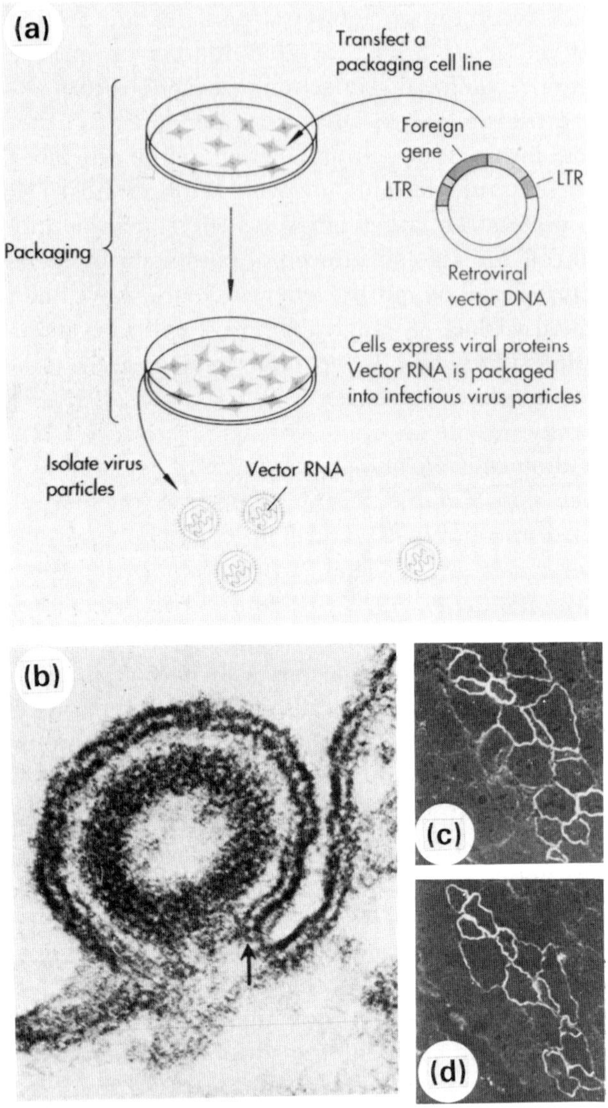

Figure 8. Production of recombinant mini-dystrophin retrovirus using packaging cell lines. (a) retroviral vector plasmid is transfected into packaging cells possessing appropriate *gag, pol* and *env* genes. The provirus (between the LTR sequences) is incorporated into infectious but replication-defective particles or virions which bud from the packaging cells (b). In (c) and (d) are shown immunostained myofibres from *mdx* muscle several weeks after intramuscular inoculation with recombinant retrovirus expressing the BMD-based dystrophin minigene.

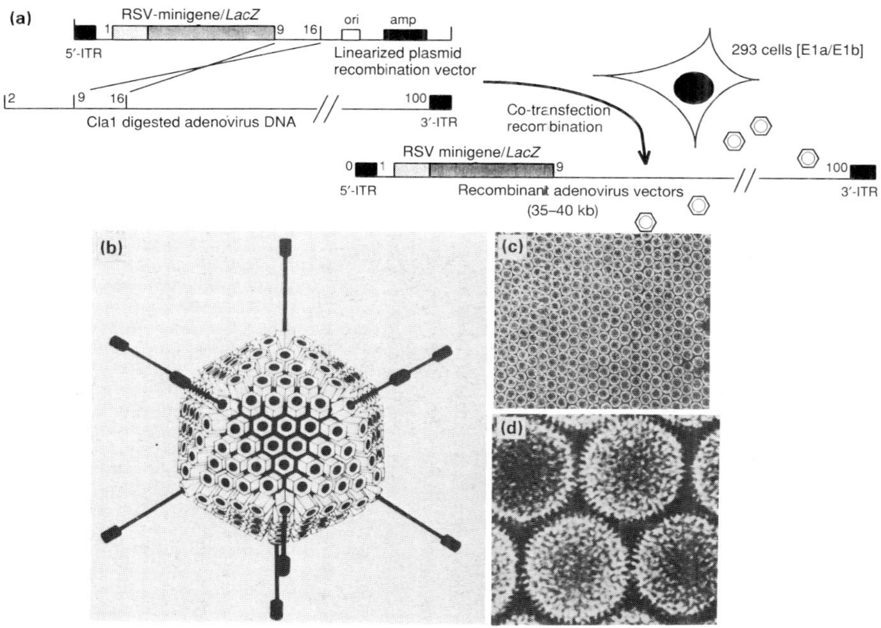

Figure 9. Adenovirus-based gene transfer vectors for dystrophin minigene expression. Plasmid shuttle vectors containing RSV-minigene constructs embedded in sequences from the left-hand end of the adenovirus-5 genome have been produced (Ragot *et al.*, 1993) and recombined in the *trans*-complementing 293 cell line with a mutant adenovirus genome (a). Replication-defective recombinant adenovirus particles are thus generated, which can be propagated in 293 cells, and can be purified to extremely high titres by $CsCl_2$-gradient centrifugation. A schematic representation of adenovirus illustrating the penton, hexon and protruding fibre structures of the capsid is shown in (b), and electron microscopic images of purified particles in (c) and (d).

obviously excludes the possibility of transferring full-length dystrophin cDNAs using these methods. However, as discussed above, based on the mild phenotype observed in some BMD patients with large in-frame deletions, the transfer of the complete gene sequence may not be necessary. As discussed above, one such mini-dystrophin gene of around 6.3 kb has been cloned into MoMLV-based retroviral and adenovirus type 5-based vectors. In the case of the retroviral vector, successful infection of cultured *mdx* satellite cells was demonstrated (Dunckley *et al.*, 1992) and direct intramuscular injection of small doses of BMD-retroviral particles into the tibialis anterior muscle of adult *mdx* mice resulted in the expression of dystrophin at the sarcolemma in approximately 6–10% of myofibres (Dunckley *et al.*, 1993; *Figure 8*). Interestingly, the expression of the BMD-based mini-dystrophin was observed for over 9 months and was found to be associated with the activated reappearance of a least one component (43 kDa) of the DAGC (Dunckley *et al.*, 1993). Experimentally induced regeneration enhanced the efficiency of this retrovirus-mediated transduction to 10–13% of fibres, suggesting that activated satellite cells are a feasible route for direct and stable dystrophin gene transfer into muscle *in vivo*. More recently, a replication-defective adenoviral vector has been engineered and used to transfer the same BMD cDNA gene into *mdx* skeletal muscle (Ragot *et al.*, 1993). Here 5–50% dystrophin-positive fibres have been reported following intramuscular injections of very large numbers of adenoviral particles ($\sim 10^9$) in neonatal mice. Furthermore, the stabilization of genetically corrected fibres was observed (Vincent *et al.*, 1993), whereas control *mdx* myofibres infected only with a β-galactosidase-expressing virus were lost, suggesting that satellite cells were not being transduced. These studies on adenoviral-mediated mini-dystrophin gene transfer in neonatal *mdx* are most encouraging. However, it remains to be determined whether long-term stability and high efficiency of gene transfer can be achieved in older animals, where inflammatory responses to adenoviral vectors have been observed (H. Davis, personal communication)

9. Future perspectives

Three main approaches have been utilized to introduce dystrophin genes into skeletal muscle, namely: (1) the transplantation of normal myoblasts (Partridge *et al.*, 1989); (2) the direct injection of expression plasmids bearing dystrophin cDNAs (Acsadi *et al.*, 1991); and (3) the transfer of functional recombinant genes by viral vector systems. All of these systems have been applied both *in vivo* and *in vitro* with varying degrees of success (Dickson *et al.*, 1991; Dunckley *et al.*, 1993; Ragot *et al.*, 1993; Vincent *et al.*, 1993; Wolff *et al.*, 1990). With heart cells, however, the only successful long-term gene transfer has been by the simple direct injection of DNA into ventricular regions *in vivo* (Acsadi *et al.*, 1991; Barr *et al.*, 1994). Unfortunately, the overall efficiency of these various procedures has been too low to offer any clinical potential.

Clinically, the intramuscular injection of gene transfer vectors is feasible only for selected muscle groups. In the future it will therefore be necessary to examine alternative systemic delivery routes for viral particles, such as those offered by intravenous or intraperitoneal administration. However, it is likely that a prolonged exposure to recombinant viral particles or repeated dosage would eventually evoke an immunological response in the host and, in these circumstances, experimental or clinical manipulation of immune surveillance might prove necessary.

Also, the potential systemic dissemination of recombinant virus raises the question of how to target gene expression to the appropriate cell types (e.g. skeletal muscle). The envelope proteins of standard retroviral vectors confer widespread cell type infectivity in most mammalian species. However, it may be possible to produce modified envelope protein specificities, by fusion to specific cell-binding ligands, such as immunoglobulins or cell adhesion molecules. In the case of the adenoviral vectors, while it remains theoretically possible that the tropism of this viral system could be specified by genetically engineering the viral coat proteins, at present this represents an immense technical challenge.

As an alternative to manipulating viral tropism, restricted transgene activity potentially could be achieved by the use of muscle-specific promoter/enhancer elements to drive expression. Large 2–6 kb fragments of the muscle creatine kinase and skeletal actin genes have been used to direct muscle-specific dystrophin expression in transgenic *mdx* mice (Cox *et al.*, 1993; Dickson *et al.*, unpublished data). Smaller regions from these types of regulatory elements and, indeed, the promoter/enhancer structures of the DMD gene itself (Barnea *et al.*, 1990) may retain appropriate tissue specificity and be sufficiently small to allow inclusion in retroviral and adenoviral dystrophin cDNA constructions. The design and production of viral gene transfer vectors allowing widespread transgene dissemination, but which possess appropriate cell type-specific tropism and/or gene transcriptional activity, is one of the most important areas of gene therapy research, and will be a vital component in the future development of gene therapeutic strategies for the treatment of DMD.

References

Acsadi G, Dickson G, Love DR, Jani A, Walsh FS, Gurusinghe A, Wolff JA and Davies KE. (1991) Human dystrophin expression in *mdx* mice after intramuscular injection of DNA constructs. *Nature,* **352**, 815–818.

Ahn AH and Kunkel LM. (1993) The structural and functional diversity of dystrophin. *Nature Genetics,* **3**, 283–291.

Bakker AJ, Head SI, Williams DA and Stephenson DG. (1993) Ca^{2+} levels in myotubes grown from the skeletal muscle of dystrophic (*mdx*) and normal mice. *J. Physiol.* **460**, 1–13.

Barnea E, Zuk D, Simantov R, Nudel U and Yaffe D. (1990) Specificity of expression of the muscle and brain dystrophin gene promoters in muscle and brain cells. *Neuron,* **5**, 881–888.

Barr E, Carroll J, Kalyncych AM, Tripathy SK, Kozarshy K, Wilson JM and Leiden JM. (1994) Efficient catheter-mediated gene transfer into the heart using replication-defective adenovirus. *Gene Ther.* **1**, 51–58.

Berkner KL. (1992) Expression of heterologous sequences in adenoviral vectors. *Curr. Top. Microbiol. Immunol.* **158**, 39–66.

Blake DJ, Tinsley JM and Davies KE. (1993) The emerging family of dystrophin-related proteins. *Trends Cell Biol.* **4**, 19–23.

Boyce FM, Beggs AH, Feener C and Kunkel LM. (1991) Dystrophin is transcribed in brain from a distant upstream promoter. *Proc. Natl Acad. Sci. USA,* **88**, 1276–1280.

Bulfield G, Siller WG, Wight PAL and Moor KJ. (1984) X chromosome-linked muscular dystrophy (*mdx*) in the mouse. *Proc. Natl Acad. Sci. USA,* **81**, 1189–1192.

Burns JC, Friedmann T, Driever W, Burrascano M and Yee JK. (1993) Vesicular stomatitis virus G glycoprotein pseudotyped retroviral vectors: concentration to very high titer and efficient gene transfer into mammalian and nonmammalian cells. *Proc. Natl Acad. Sci. USA,* **90**, 8033–8037.

Byers TJ, Kunkel LM and Watkins SC. (1991) The subcellular distribution of dystrophin in mouse skeletal, cardiac and smooth muscle. *J. Cell Biol.* **115**, 411–421.

Campion DR. (1984) The muscle satellite cell: a review. *Int. Rev. Cytol.* **87**, 225–251.

Clerk A, Morris GE, Dubowitz V, Davies K and Sewry CA. (1993) Dystrophin-related protein, utrophin, in normal and dystrophic human fetal skeletal muscle. *Histochem. J.* **25**, 554–561.

Cox GA, Cole NM, Matsumura K, Phelps SF, Hauschka SD, Campbell KP, Faulkner JA and Chamberlain JS. (1993) Over expression of dystrophin in transgenic *mdx* mice eliminates dystrophic symptoms without toxicity. *Nature,* **364**, 725–729.

Davis HL, Demeneix BA, Quantin B, Coulombe J and Whalen RG. (1993)

Plasmid DNA is superior to viral vectors for direct gene transfer into adult mouse skeletal muscle. *Hum. Gene Ther.* **4**, 733–740.

Den Dunnen JT, Grootscholten PM, Dauwerse JG *et al.* (1992) Reconstruction of the 2.4 Mb human DMD-gene by homologous YAC recombination. *Hum. Mol. Genet.* **1**, 19–28.

Dickson G and Dunckley MG. (1993) Human dystrophin gene transfer: genetic correction of dystrophin deficiency. In: *Molecular and Cell Biology of Muscular Dystrophy* (ed. T. Partridge). Chapman & Hall, London, pp. 283–302.

Dickson G, Love DR, Davies KE, Wells KE, Piper TA and Walsh FS. (1991) Human dystrophin gene transfer: production and expression of a functional recombinant DNA-based gene. *Hum. Genet.* **88**, 53–58.

Dickson G, Azad A, Morris GE, Simon H, Noursadeghi M and Walsh FS. (1992) Co-localization and molecular association of dystrophin with laminin at the surface of mouse and human myotubes. *J. Cell Sci.* **103**, 1223–1233.

Dunckley MG and Dickson G. (1994) Retroviral-mediated gene transfer and Duchenne muscular dystrophy. In: *Gene Therapeutics* (ed. J.A. Wolff). Birkhauser, Boston, MA, pp. 391–410.

Dunckley MG, Love DR, Davies KE, Walsh FS, Morris GE and Dickson G. (1992) Retroviral-mediated transfer of a dystrophin minigene into *mdx* myoblasts *in vitro*. *FEBS Lett.* **296**, 128–134.

Dunckley MG, Wells DJ, Walsh FS and Dickson G. (1993) Direct retroviral-mediated transfer of a dystrophin minigene into *mdx* mouse muscle *in vivo*. *Hum. Mol. Genet.* **2**, 717–723.

Eglitis MA, Kantoff P, Gilboa E and Anderson WF. (1985) Gene expression in mice after high efficiency retroviral-mediated gene transfer. *Science*, **230**, 1395–1398.

Emery AEH. (1993) *Duchenne Muscular Dystrophy*, 2nd edn. Oxford University Press, Oxford.

Engel AG. (1993) Gene therapy for Duchenne dystrophy. *Ann. Neurol.* **34**, 3–4.

England SB, Nicholson LVB, Johnson MA, Forrest SM, Love DR, Zubrzycka-Gaarn EE, Bulman DE, Harris JB and Davies KE. (1990) Very mild muscular dystrophy associated with deletion of 46% of dystrophin. *Nature*, **343**, 180–182.

Ervasti JM and Campbell KP. (1991) Membrane organization of the dystrophin–glycoprotein complex. *Cell*, **66**, 1121–1131.

Feener CA, Koenig M and Kunkel LM. (1989) Alternative splicing of human dystrophin mRNA generates isoforms at the carboxy terminus. *Nature*, **338**, 509–511.

Franco A and Lansman JB. (1990) Calcium entry through stretch-inactivated ion channels in *mdx* myotubes. *Nature*, **344**, 670–673.

Gaschen FP, Hoffman EP, Gorospe JR, Uhl EW, Senior DF, Cardinet GH and Pearce LK. (1992) Dystrophin deficiency causes lethal muscle hypertrophy in cats. *J. Neurol. Sci.* **110**, 149–159.

Gnirke A and Huxley C. (1991) Transfer of the human HPRT and GART genes from yeast to mammalian cells by microinjection of YAC DNA. *Som. Cell Mol. Genet.* **17**, 573–580.

Gnirke A, Barnes TS, Patterson D, Schild D, Featherstone T and Olson MV. (1991) Cloning and *in vivo* expression of the human GART gene using yeast artificial chromosomes. *EMBO J.* **10**, 1629–1634.

Graessmann M and Graessmann A. (1983) Microinjection of tissue culture cells. *Methods Enzymol.* **101**, 482–492.

Hemmings L, Kuhlman PA and Critchley DR. (1992) Analysis of the actin-binding domain of α-actinin by mutagenesis and demonstration that dystrophin contains a functionally homologous domain. *J. Cell Biol.* **116**, 1369–1380.

Hoffman EP and Kunkel LM. (1989) Dystrophin abnormalities in Duchenne/Becker muscular dystrophy. *Neuron,* **2**, 1019–1029.

Hoffman EP, Brown RH and Kunkel LM. (1987) Dystrophin: the protein product of the Duchenne muscular dystrophy locus. *Cell,* **51**, 919–928.

Ibraghimov-Beskrovnaya O, Ervasti JM, Leveille CJ, Slaughter CA, Sernett SW and Campbell KP. (1992) Primary structure of dystrophin-associated glycoproteins linking dystrophin to the extracellular matrix. *Nature,* **355**, 696–702.

Karpati G, Carpenter S, Morris GE, Davies KE, Guerin C and Holland P. (1993a) Localization and quantitation of the chromosome 6-encoded dystrophin-related protein in normal and pathological human muscle. *J. Neuropathol. Exp. Neurol.* **52**, 119–128.

Karpati G, Adjukovic D, Arnold D *et al.* (1993b) Myoblast transfer in Duchenne muscular dystrophy. *Ann. Neurol.* **34**, 8–17.

Koenig M, Monaco AP and Kunkel LM. (1988) The complete sequence of dystrophin predicts a rod-shaped cytoskeletal protein. *Cell,* **53**, 219–228.

Koenig M, Beggs AH, Moyer M *et al.* (1989) The molecular basis for Duchenne versus Becker muscular dystrophy: correlation of severity with type of deletion. *Am. J. Hum. Genet.* **45**, 498–506.

Lee CC, Pearlman JA, Chamberlain JS and Caskey CT. (1991) Expression of recombinant dystrophin and its localization to the cell membrane. *Nature,* **349**, 334–336.

Love DR, Blyth BC, Tinsley JM, Blake DJ and Davies KE. (1993) Dystrophin and dystrophin-related proteins: a review of protein and RNA studies. *Neuromusc. Disord.* **3**, 5–21.

Matsumura K, Ervasti JM, Ohlendieck K, Kahl SD and Campbell KP. (1992a) Association of dystrophin-related protein with dystrophin-associated proteins in *mdx* mouse muscle. *Nature,* **360**, 588–591.

Matsumura K, Tome FMS, Collin H, Azibi K, Chaouch M, Kaplan JC, Fardeau M and Campbell KP. (1992b) Deficiency of the 50K dystrophin-associated glycoprotein in severe childhood autosomal recesssive muscular dystrophy. *Nature,* **359**, 320–322.

Menke A and Jockusch H. (1991) Decreased osmotic stability of dystrophin-less muscle cells from the *mdx* mouse. *Nature,* **349**, 69–71.

Miller AD. (1992) Human gene therapy comes of age. *Nature,* **357**, 455–460.

Monaco AP, Bertelson CJ, Liechti-Gallati S, Moser H and Kunkel LM. (1988) An explanation for the phenotypic differences between patients

bearing partial deletions of the DMD locus. *Genomics*, **2**, 90–95.
Mulligan RC. (1993) The basic science of gene therapy. *Science*, **260**, 926–931.
Noursadeghi M, Walsh FS, Heiman-Patterson T and Dickson G. (1993) Trans-activation of the murine dystrophin gene in human–mouse hybrid myotubes. *FEBS Lett.* **320**, 155–159.
Ohlendieck K and Campbell KP. (1991) Dystrophin constitutes 5% of membrane cytoskeleton in skeletal muscle. *FEBS Lett.* **283**, 230–234.
Ohlendieck K, Ervasti JM, Matsumura K *et al.* (1991) Dystrophin related protein is localised to neuromuscular junctions of adult skeletal muscle. *Neuron*, **7**, 499–508.
Partridge TA, Morgan JE, Coulton GR, Hoffman EP and Kunkel LM. (1989) Conversion of *mdx* myofibers from dystrophin-negative to -positive by injection of normal myoblasts. *Nature*, **337**, 176–179.
Pavlath GK, Rich K, Webster SG and Blau HM. (1989) Localization of muscle gene products in nuclear domains. *Nature*, **337**, 570–573.
Porter GA, Dmytrenko GM, Winkelmann JC and Block RJ. (1992) Dystrophin colocalises with β-spectrin in distinct subsarcolemmal domains in mammalian skeletal muscle. *J. Cell Biol.* **117**, 997–1005.
Poznansky M, Lever A, Bergeron L, Haseltine W and Sodroski J. (1991) Gene transfer into human lymphocytes by a defective human immunodeficiency virus type 1 vector. *J. Virol.* **65**, 532–536.
Quantin B, Perricaudet LD, Tajbakhsh S and Mandel JL. (1992) Adenovirus as an expression vector in muscle cells *in vivo*. *Proc. Natl Acad. Sci. USA*, **89**, 2581–2584.
Ragot T, Vincent N, Chafey P. *et al.* (1993) Efficient adenovirus mediated transfer of a human minidystrophin gene to skeletal muscle of *mdx* mice. *Nature*, **361**, 647–650.
Robards SL, Ervasti JM, Anderson RD, Ohlendieck K, Kahl SD, Zoloto D and Campbell KP. (1993) Disruption of the dystrophin–glycoprotein complex in the cardiomyopathic hamster. *J. Biol. Chem.* **268**, 11496–11499.
Roberts RG, Bobrow R and Bentley DR. (1992) Point mutations in the dystrophin gene. *Proc. Natl Acad. Sci. USA*, **89**, 2331–2335.
Roberts RG, Coffey AJ, Bobrow M and Bentley DR. (1993) Exon structure of the human dystrophin gene. *Genomics*, **16**, 536–538.
Rosenfeld MA, Yoshimura K, Trapnell BC *et al.* (1992) *In vivo* transfer of human cystic fibrosis transmembrane conductance regulator gene to the airway epithelium. *Cell*, **68**, 143–155.
Sewry CA, Sansome A, Clerk A, Sherratt TG, Hasson N, Rodillo E, Heckmatt JZ, Strong PN and Dubowitz V. (1993) Manifesting carriers of Xp21 muscular dystrophy; lack of correlation between dystrophin expression and clinical weakness. *Neuromus. Disord.* **3**, 141–148.
Smith TG, Mehaffey MG, Kayda DB, Saunders JM, Yei S, Trapnell BC, McClelland A and Kaleko M. (1993) Adenovirus mediated expression of therapeutic plasma levels of human factor IX in mice. *Nature Genetics*, **5**, 397–402.
Stedman HH, Sweeney HL, Shrager JB *et al.* (1991) The *mdx* mouse

diaphragm reproduces the degenerative changes of Duchenne muscular dystrophy. *Nature*, **352**, 536–538.

Stratford-Perricaudet LD, Makeh I, Perricaudet M and Briand P. (1992) Widespread long-term gene transfer to mouse skeletal muscles and heart. *J. Clin. Invest.* **90**, 626–630.

Strauss WM and Jaenisch R. (1992) Molecular complementation of a collagen mutation in mammalian cells using yeast artificial chromosomes. *EMBO J.* **11**, 417–421.

Suzuki A, Yoshida M, Yamamoto H and Ozawa E. (1992) Glycoprotein-binding site of dystrophin is confined to the cysteine-rich domain and the first half of the carboxy-terminal domain. *FEBS Lett.* **308**, 154–160.

Turner PR, Fong PY, Denetclaw WF and Steinhardt RA. (1991) Increased calcium influx in dystrophic muscle. *J. Cell Biol.* **115**, 1701–1712.

Valentine BA, Cooper BJ, Cummings JF and DeLahunta A. (1990) Canine X-linked muscular dystrophy: morphologic lesions. *J. Neurol. Sci.* **97**, 1–23.

Vincent N, Ragot T, Gilgencrantz H *et al.* (1993) Long-term correction of mouse dystrophic degeneration by adenovirus-mediated transfer of a minidystrophin gene. *Nature Genetics*, **5**, 130–134.

Walsh FS, Pizzey JA and Dickson G. (1989) Tissue-specific isoforms of dystrophin. *Trends Neurosci.* **12**, 235–238.

Weller B, Karpati G and Carpenter S. (1990) Dystrophin-deficient *mdx* muscle fibers are preferentially vulnerable to necrosis induced by experimental lengthening contractions. *J. Neurol. Sci.* **100**, 9–15.

Wells DJ. (1993) Improved gene transfer by direct plasmid injection associated with regeneration in mouse skeletal muscle. *FEBS Lett.* **332**, 179–182.

Wells DJ and Goldspink G. (1992) Age and sex influence expression of plasmid DNA directly injected into mouse skeletal muscle. *FEBS Lett.* **306**, 203–205.

Wells DJ, Wells KE, Walsh FS *et al.* (1992) Human dystrophin expression corrects the myopathic phenotype in transgenic *mdx* mice. *Hum. Mol. Genet.* **1**, 35–40.

Wessel HB. (1990) Dystrophin: a clinical perspective. *Ped. Neurol.* **6**, 3–12.

Williams RS, Johnston SA, Riedy M *et al.* (1991). Introduction of foreign genes into tissues of living mice by DNA-coated microprojectiles. *Proc. Natl Acad. Sci. USA*, **88**, 2726–2730.

Wolff JA, Malone RW, Williams P, Chong W, Acsadi G, Jani A and Felgner PL. (1990) Direct gene transfer into mouse muscle *in vivo*. *Science*, **247**, 1465–1468.

Wolff JA, Williams P, Acsadi G, Jiao S, Jani A and Chong W. (1991) Conditions affecting direct gene transfer into rodent muscle *in vivo*. *Biotechniques*, **11**, 474–485.

Zabner J, Couture LA, Gregory RA *et al.* (1993) Adenovirus-mediated gene transfer transiently corrects the chloride transport defect in nasal epithelia of patients with cystic fibrosis. *Cell*, **75**, 207–216.

Zhu N, Liggit D, Liu Y and Debs R. (1993) Systemic gene expression after intravenous DNA delivery into adult mice. *Science*, **261**, 209–211.

5

Molecular genetics of leukaemia

M.F. Greaves
Leukaemia Research Fund Centre, Institute of Cancer Research, Chester Beatty Laboratories, London, UK

1. Introduction: clonal evolution of cancer cells

Leukaemia is a blood cell cancer and, in essence, a genetic disorder but, in contrast to most of the other genetic diseases discussed in this volume, almost all of the mutations involved are acquired and not inherited. These mutations arise as rare events in a single somatic cell and, as a consequence, leukaemia is a *clonal* disorder (Greaves, 1991), as are other cancers. Also in common with other cancers, the great majority of leukaemias have more than one mutation within a clone. It appears that, as in colon and breast carcinomas, a succession of sequential mutations occur that are complementary or synergistic in their functional impact. There are a remarkable number of different genes involved in leukaemia but they have, in principle, a restricted range of primary functions as mutants: dysregulation of proliferation, differentiation, cell death or repair of DNA damage. Each gene that is involved has selective advantage for the clone for the single reason that it directly or indirectly endows the clone with *net reproductive advantage* as a mutant. This is an important characteristic and emphasizes the analogy that has been drawn before between the evolution of malignant clones and the evolution of new species from germline mutants.

To pursue the Darwinian analogy further, it is apparent that what happens in leukaemia is that, as a result of the emergence of the leukaemic clone, there is competition with normal clones in the bone marrow for space and nutrients. Normal function is therefore compromised, which results in a deficiency of red cells, platelets and granulocytes. This gives rise, in turn, to essentially all the pathology of the disease, clinical symptoms and potentially lethal consequences.

Janet Rowley was perhaps the first to emphasize that in different subtypes of myeloid leukaemia (and subsequently in other blood cell cancers) there are different non-random chromosome translocations (Heim and Mitelman, 1987; Rowley, 1984). This extremely influential insight led to the implications that, first, because these changes were non-random it was likely that the genetic changes were causally relevant and, second, it was speculated that the break points for the chromosome changes provided landmarks for the genes involved. This has turned out to be absolutely correct. Fortunately, at about the time when these data emerged, 10 years ago, molecular technology was also becoming available to pursue these predictions. Another crucial handle on these genes was provided by molecular virology studies and soon several genes were cloned at these chromosome break points, starting with c-*myc* (Dalla-Favera *et al.*, 1982). More recently, some of the cloned genes have been transfected into fertilized eggs to make transgenic animals and those animals have come down with the 'appropriate' type of leukaemia, satisfying Koch's postulate for the causal role of these genes (Adams and Cory, 1991).

An equally important correlate of these discoveries has been the clinical issue of whether particular translocations and the underlying mutations are associated with the patients' response to treatment. This may, of course, depend on the particular gene and on the therapy applied; indeed, as therapies improve some of the associations identified earlier have disappeared! But clearly the presence or absence of these mutations can be clinically important. The mutation that has been best studied in this respect is the Philadelphia (Ph) chromosome and the underlying *bcr–abl* fusion gene. Combination chemotherapy and radiation, which has been extremely effective for some leukaemias (especially in children), seldom salvage patients with Ph-positive acute leukaemia (Secker-Walker *et al.*, 1991). Why particular molecular abnormalities should impart clinical unresponsiveness is not fully understood, but important factors appear to be the intrinsic drug sensitivity of the cell type in which the mutation arises and the ability of many oncogene mutations (e.g. in *bcl-2, p53, bcr–abl* and *ras*) to render cells less vulnerable to drug- and radiation-induced cell death by apoptosis (Dive and Hickman, 1991; Eastman, 1990; Greaves, 1993). It is critical to identify patients whose leukaemic clone has a genetic abnormality, such as the Ph chromosome, some of whom may be cured by bone marrow transplantation. It follows that molecular abnormalities provide important markers for differential diagnosis of the disease, for monitoring the response to treatment (i.e. detection of residual leukaemic cells with polymerase chain reaction (PCR); Negrin and Blume, 1991; Yokota *et al.*, 1991), and potential targets for molecular therapy. The latter may, in the long term, be the most important application of this knowledge.

2. Molecular diversity of mutations in leukaemia

The molecular genetics of leukaemia have revealed an extraordinary diversity of abnormalities. This heterogeneity can be demonstrated by contrasting two different leukaemias, acute lymphoblastic leukaemia (ALL), which is the major paediatric cancer in the Western world, and chronic myeloid leukaemia (CML), which is a common leukaemia subtype in adults. ALL is a group of malignant blood cell cancers originating primarily in lymphocyte progenitors but also (in adults at least) in lymphomyeloid stem cells. It is characterized by a striking variety of molecular abnormalities (Berger, 1992; Pui *et al.*, 1990) (*Table 1*). Adding up the number of genes that may be involved in this disease gives a total of more than 40 non-random translocations, a number of deletions and point mutations, plus, presumably, other mutations that we have not yet discovered. The most striking recent finding in the molecular genetics of ALL has been the discovery that many of the genes disrupted and/or dysregulated as a result of chromosome translocations encode transcription factors (Cleary, 1991). The total number of genes involved in all types of leukaemias is unknown but it is thought to be approximately 200. If there are five times more genes involved in cancer in total then 1% of our 10^5 genes are potential oncogenes. This is not a surprising number if these same genes are critical in proliferation, differentiation and cell death.

The story is rather different with CML which, at diagnosis, is a more homogeneous and benign leukaemia originating in lymphomyeloid stem cells. As we showed some years ago using molecular probes (Chan *et al.*, 1987), all CML, irrespective of genetic background or geographic region, involves essentially the same mutation: a fusion between *bcr* (22q11) and *abl* (9q34). Although CML is initially benign, it inevitably progresses to an acute leukaemia or blast crisis after an average of about 3 years. A whole gamut of different mutations is observed when CML becomes an acute leukaemia in differentiation arrest (blast crisis) (Heim and Mitelman, 1987). This scenario of divergent genetic evolution is a common theme in cancers, and in this respect CML in blast crisis is, in molecular terms,

Table 1. Common genetic abnormalities in acute lymphoblastic leukaemia

Type			Example
Translocations	(a)	Dysregulation by juxtaposition with IgH or TCR (α, β, δ) genes	*myc*, *hox*11/*rmt*1/2, *lyl*1
	(b)	Novel (hybrid) product by in-frame gene fusion	*bcr–abl* *pbx*–E2A *hrx* (or *mll*) – with multiple partners
Deletions	9p$^-$	(~30%)	
Point mutations	N-*ras*	(~20%)	
Hyperdiploidy	(Chromosome 21 + others)	(15%)	

equivalent to ALL (or acute myeloblastic leukaemia (AML)).

So why so many genes? There appear to be parallel pathways of sequential mutations reflecting both molecular co-operativity in transformation and functional redundancy in the system, different subtypes of leukaemia arising (in part) from mutations in different genes and pathways, and genetic divergence as the disease progresses and becomes more malignant (*Figure 1*). From a biological point, this perspective accommodates much of the molecular diversity observed but it is important to emphasize that this variability makes for a severe clinical challenge in terms of selective diagnosis and effective therapy.

3. Diagnostic applications of molecular markers

There is an embarrassment of riches in the application of these techniques because of the many different ways in which these mutations can now be detected. For example, with respect to the Ph chromosome (*bcr–abl* gene fusion) the abnormality can be detected by Southern blot, by pulse field gel electrophoresis, the mRNA product can be detected by Northern blots or by reverse PCR, the fusion protein can be detected by enhanced kinase activity and, more recently, antibodies (Morgan and Wiedemann, 1992). PCR is probably the method of choice for detecting residual disease in treated patients, i.e. detecting approxi-

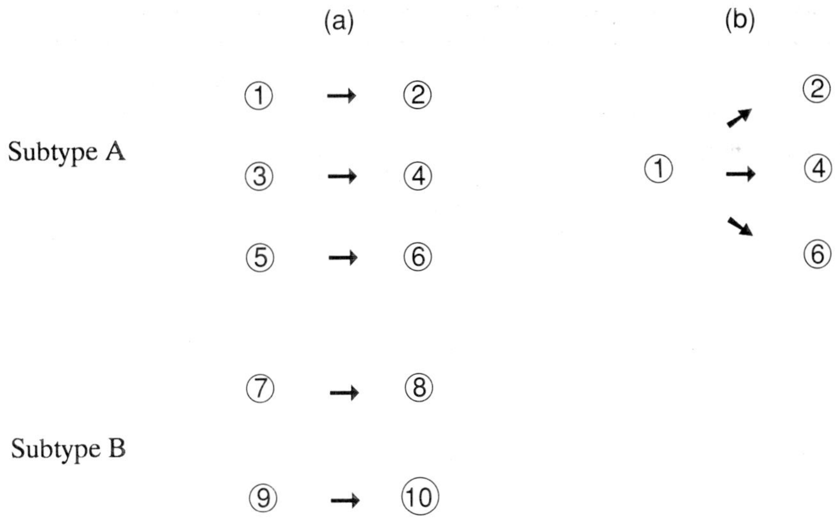

Figure 1. Mutation and clonal evolution in leukaemia. (a) Parallel pathways, (b) divergent pathways.

mately one cell in 10^5 (Cross *et al.*, 1993; Morgan *et al.*, 1989), but it is not a sufficiently robust technique for routine diagnosis. A more practical method is required that can be applied to ordinary blood smears and that can be carried out quickly. The recently developed fluorescence *in situ* hybridization (FISH) technology, which can be applied to non-dividing (interphase) cells with molecular abnormalities (Gray and Pinkel, 1992), is the best option available, and it is likely to have a major impact in molecular medicine.

The technique would be most useful for the leukaemias if it could be applied to blood or bone marrow smears and combined with immunological analysis, in order to tell us at the same time what sort of cell we were looking at. We have recently developed a dual staining technique which allows this to be done very effectively (Price *et al.*, 1992). The method works particularly well using chromosome-specific centromeric repeat probes to identify triploidy or chromosome loss, but it should also be possible to detect translocations using yeast artificial chromosome (YAC) or other large clones that span the break points (Lengauer *et al.*, 1992). Classical chromosome analysis will continue to have a role to play as a front line genetic screening procedure but it is important to recognize that, in several instances, molecular probing has revealed a higher frequency of particular abnormalities than was estimated from karyotypic studies, e.g. *tal* gene deletions and rearrangements in T-ALL, 9p deletions in ALL, *bcr–abl* gene fusions in so-called 'Ph-negative' CML, *hrx* (*mll*) gene rearrangements in infant ALL.

4. Therapeutic possibilities: targeting oncogenes

Several potential methods are available for treating cancer by targeting oncogenes and their products, although these ideas are challenging and difficult to translate into practice. One strategy is the immunological route using antibodies (Roy *et al.*, 1991) or cytotoxic T-cell clones (Chen *et al.*, 1992; Yanuck *et al.*, 1993). It is fascinating that immunotherapy, which has had a rather chequered history in cancer treatment, is now being resurrected in a much more credible way, and we can anticipate some exciting developments now that we know which molecules to aim at. A second strategy is to use substrate or receptor antagonists or ligands to inhibit enzyme (e.g. kinase) activity or to block receptors (Levitzki and Gilon, 1991). The third approach that may have particular applications in leukaemia is by using strategies for inhibiting expression of the offending mutant gene – either at the DNA level itself or at the mRNA level. One of the approaches being studied intensively in this respect is the use of anti-sense oligonucleotides that either inhibit mRNA translation or induce its degradation (Ratajczak *et al.*, 1992; Stein, 1992). Few believe that this is going to work therapeutically with a simple shot of oligo. Important developments will be to modify these compounds chemically by altering or replacing the DNA backbone to increase stability whilst not sacrificing specificity and cell uptake and, perhaps, to introduce these compounds into expression systems so that there is stable inhibition of gene expression. If several of the mutant genes in leukaemia have a powerful anti-apoptotic effect, then even temporary and partial inhibition of gene expression might render the clone much more drug-responsive (Miyashita and Reed, 1992). The very considerable challenge in this field is primarily in the chemistry rather than the biology. The reason why leukaemia is such an attractive cancer to tackle in this respect is that we have some specific target mRNAs to aim at: hybrid genes with unique junctional sequences (*Table* 2).

5. Oncogene mutations and aetiology

An insight into the mutations involved in leukaemia may help us to understand the molecular pathogenesis of these diseases in the rather broader context of their natural history and aetiology. In one particular

Table 2. Hybrid gene mRNAs provide leukaemia-specific targets for anti-sense oligonucleotides

Fused genes	Chromosome translocation	Leukaemia type
bcr–abl	(Ph)t(9;22)(q34;q11)	CML/AL
pbx–e2a	t(1;19)(q23;p13)	pre-B(μ^+) ALL
dek–can	t(6;9)(p23;q34)	AML (M2/M4 baso)
pml–rarα	t(15;17)(p22;q11/12)	APML (M3)

CML, chronic myeloid leukaemia; AL, acute leukaemia (lymphoid or myeloid; ALL, acute lymphoblastic leukaemia; AML, acute myeloblastic leukaemia; APML, acute promyelocytic leukaemia; M2, M3, M4, morphological classification of subtype; pre-B(μ^+), B-cell precursor with cytoplasmic μ heavy chain of immunoglobulin; M4 baso, M4 morphological subtype of AML with basophils.

type of leukaemia we have been able to identify the developmental timing of a critical leukaemogenic mutation. The majority (~75%) of infant (<12 months) acute leukaemias have a consistent involvement of chromosome 11q23 in which there is a reciprocal translocation with a remarkable variety of other chromosomes and genes (Pui *et al.*, 1990; Thirman *et al.*, 1993). The relevant gene, located at 11q23, has recently been cloned and bears homology to the *Drosophila* developmental gene *trithorax* (Tkachuk *et al.*, 1992). Rearrangements of this gene, called *hrx* (or *mll*), occur in introns and cluster in a region of approximately 10 kb, individual patients having different break points (Morgan *et al.*, 1992). Recently, we have exploited this information in the context of leukaemia in identical twins to demonstrate that the *hrx* mutation, when found in infant acute leukaemias, is likely to have arisen during pregnancy (Ford *et al.*, 1993).

The concordance rate for paediatric leukaemia in monozygous (identical) twins is approximately 25% (Zuelzer and Cox, 1969). This could be explained by inherited susceptibility, with a one in four chance that both twins will develop leukaemia. The alternative possibility, suggested some 20 years ago, is that a single leukaemogenic transformation *in utero* in one twin could lead to the establishment of the malignant clone in the other twin by metastasis via a shared placental circulation (Clarkson and Boyse, 1971).

If twins with concordant leukaemia were to share a unique but non-constitutive oncogene rearrangement in their leukaemic cells, then this would provide evidence both for an *in utero* origin of that leukaemogenic event and unequivocal support for the 'metastasis' hypothesis for leukaemia concordance in twins. The recent cloning of the *hrx* gene at 11q23 allowed this idea to be tested.

The introns within which patient-specific breaks occur at 11q23 in infant ALL have yet to be sequenced, but identity or non-identity of *hrx* rearrangements will be evident from restriction fragment size in Southern blots. We analysed three pairs of infant twin leukaemic cell DNAs for *hrx* rearrangements by Southern blotting and found that all three pairs had *hrx* rearrangements. Restriction fragments from DNA of paired twin leukaemia samples co-migrated but they differed between twin pairs and were also different from 10 other controls of non-twinned infant ALL with 11q23 translocations (Ford *et al.*, 1993). Data from one pair of twins and three controls are illustrated in *Figure 2*.

These data provide strong evidence that the rearrangements in paired twins are identical and that this is unlikely to be due to chance. Evidence that the twin rearrangements detected were not inherited as structural abnormalities or restriction site polymorphisms was pro-

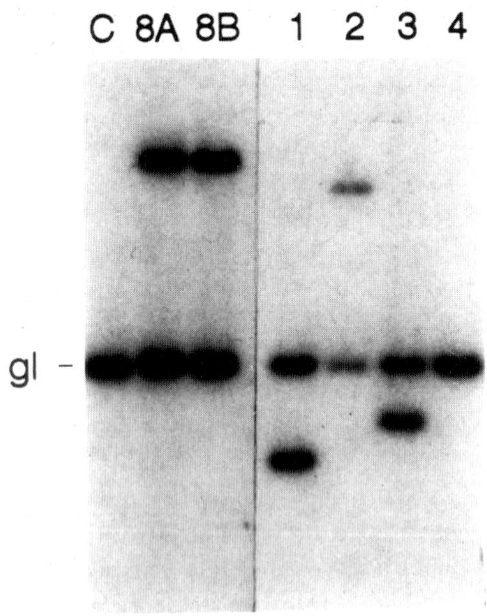

Figure 2. Rearrangement of the *trithorax*-related gene (*hrx*) in identical infant twins with ALL. Southern blot of *hrx* rearrangements. C, control; 8A, 8B, pair of identical infant twins with ALL; 1, 2 and 3, other unrelated infants with infant ALL and 11q23 translocations; 4, ALL with no 11q23 translocation.

vided by the observation that no *hrx* rearrangements were observed in non-leukaemic DNA obtained from the same twins or parental DNA. Rearrangements of *hrx* are therefore acquired and not inherited. Since the rearrangement that is shared by each twin pair is both unique and acquired, the explanation must be that it arose as a single clonal event *in utero* in one twin, generating leukaemic or pre-leukaemic clonogenic progeny. These then metastasized to the other twin via vascular anastomoses in their shared monochorionic placenta. Further evidence for a common clonal origin was provided by analysis of IgH gene rearrangements and patterns of X chromosome inactivation (Ford *et al.*, 1993).

Twins are no more at risk of developing leukaemia (as a pair) than non-twinned individuals. We draw the conclusion therefore that, in the majority of cases of *non-twinned* infant acute leukaemia, translocations that disrupt *hrx* occur during pregnancy. These findings have obvious aetiological and epidemiological implications. In particular, they focus attention on possible aetiological events or exposures during pregnancy. The identification of these events requires epidemiological investigation. A very extensive and comprehensive case/control study of the epidemiology of childhood leukaemia (and other cancers) has been initiated in the UK. It is likely that molecular analysis of leukaemia cells will make an important contribution to this and other similar studies aimed at uncovering the causes of leukaemia.

Acknowledgements

The author is supported by the Leukaemia Research Fund of Great Britain and the Kay Kendall Leukaemia Fund. Sarah Macnaghten and Barbara Deverson are thanked for help with preparation of the manuscript.

References

Adams JM and Cory S. (1991) Transgenic models of tumor development. *Science*, **254**, 1161–1167.

Berger R. (1992) The cytogenetics of haematological malignancies. In: *Baillière's Clinical Haematology* (ed. B Young). Baillière Tindall, London, pp. 791–814.

Chan LC, Chen PM, Powles R, Saragas E, Wiedemann LM, Groffen J and Greaves MF. (1987) The molecular lesion in chronic granulocytic leukaemia is highly conserved despite ethnic and geographical variation. *Leukemia*, **1**, 486–490.

Chen W, Peace DJ, Rovira DK, You S-G and Cheever MA. (1992) T-cell immunity to the joining region of $p210^{BCR-ABL}$ protein. *Proc. Natl Acad. Sci. USA*, **89**, 1468–1472.

Clarkson BD and Boyse EA. (1971) Possible explanation of the high concordance for acute leukaemia in monozygous twins. *Lancet*, **i**, 699–701.

Cleary ML. (1991) Oncogenic conversion of transcription factors by chromosomal translocations. *Cell*, **66**, 619–622.

Cross NCP, Feng L, Chase A, Bungey J, Hughes TP and Goldman JM. (1993) Competitive polymerase chain reaction to estimate the number of BCR–ABL transcripts in chronic myeloid leukemia patients after bone marrow transplantation. *Blood*, **82**, 1929–1936.

Dalla-Favera R, Bregni M, Erickson J, Patterson D, Gallo R and Croce CM. (1982) Human c-*myc* oncogene is located on the region of chromosome 8 that is translocated in Burkitt's lymphoma cells. *Proc. Natl Acad. Sci. USA*, **79**, 7824–7827.

Dive C and Hickman JA. (1991) Drug–target interactions: only the first step in the commitment to a programmed cell death. *Br. J. Cancer*, **64**, 192–196.

Eastman A. (1990) Activation of programmed cell death by anticancer agents: cisplatin as a model system. *Cancer Cells*, **2**, 275–280.

Ford AM, Ridge SA, Cabrera ME, Mahmoud H, Steel CM, Chan LC and Greaves MF. (1993) *In utero* rearrangements in the *trithorax*-related oncogene in infant leukaemias. *Nature*, **363**, 358–360.

Gray JW and Pinkel D. (1992) Molecular cytogenetics in human cancer diagnosis. *Cancer*, **69**, 1536–1541.

Greaves MF. (1991) Biology of human leukaemia. In: *Introduction to the Cellular and Molecular Biology of Cancer* (eds LM Franks and NM Teich). Oxford University Press, Oxford, pp. 330–356.

Greaves MF. (1993) Stem cell origins of leukaemia and curability. *Br. J. Cancer*, **67**, 413–423.

Heim S and Mitelman F. (1987) *Cancer Cytogenetics*. Alan R. Liss, New York.

Lengauer C, Reithman HC, Speicher MR, Taniwaki M, Konecki D, Green ED, Becher R, Olson MV and Cremer T. (1992) Metaphase and interphase cytogenetics with Alu-PCR amplified yeast artificial chromosome clones containing the BCR gene and the protooncogenes c-*raf-1*, c-*fms* and c-*erbB-2*[1] . *Cancer Res.* **52**, 2590–2596.

Levitzki A and Gilon C. (1991) Tyrphostins as molecular tools and potential antiproliferative drugs. *Trends Pharmacol. Sci.* **12**, 171–174.

Miyashita T and Reed JC. (1992) *bcl*-2 gene transfer increases relative resistance of S49.1 and WEHI7.2 lymphoid cells to cell death and DNA fragmentation induced by glucocorticoids and multiple chemotherapeutic drugs. *Cancer Res.* **52**, 5407–5411.

Morgan GJ and Wiedemann LM. (1992) Annotation: the clinical application of molecular techniques in Philadelphia-positive leukaemia. *Br. J. Haematol.* **80**, 1–5.

Morgan GJ, Hughes T, Janssen JWG, Gow J, Guo A-P, Goldman JM,

Wiedemann LM and Bartram CR. (1989) Polymerase chain reaction for detection of residual leukaemia. *Lancet*, **i**, 928–929.

Morgan GJ, Cotter F, Katz FE, Ridge SA, Domer P, Korsmeyer S and Wiedemann LM. (1992) Breakpoints at 11q23 in infant leukemias with the t(11;19)(q23;p13) are clustered. *Blood*, **80**, 2172–2175.

Negrin RS and Blume KG. (1991) The use of the polymerase chain reaction for the detection of minimal residual malignant disease. *Blood*, **78**, 255–258.

Price CM, Kanfer EJ, Colman SM, Westwood N, Barrett AJ and Greaves MF. (1992) Simultaneous genotypic and immunophenotypic analysis of interphase cells using dual-color fluorescence: a demonstration of lineage involvement in polycythemia vera. *Blood*, **80**, 1033–1038.

Pui C-H, Crist WM and Look AT. (1990) Biology and clinical significance of cytogenetic abnormalities in childhood acute lymphoblastic leukemia. *Blood*, **76**, 1449–1463.

Ratajczak MZ, Kant JA, Luger SM, Hijiya N, Zhang J, Zon G and Gewirtz AM. (1992) *In vivo* treatment of human leukemia in a *scid* mouse model with c-*myb* antisense oligodeoxynucleotides. *Proc. Natl Acad. Sci. USA*, **89**, 11823–11827.

Rowley JD. (1984) Biological implications of consistent chromosome rearrangements in leukemia and lymphoma. *Cancer Res.* **44**, 3159–3168.

Roy DC, Griffin JD, Belvin M, Blättler WA, Lambert JM and Ritz J. (1991) Anti-MY9-blocked-ricin: an immunotoxin for selective targeting of acute myeloid leukemia cells. *Blood*, **77**, 2404–2412.

Secker-Walker LM, Craig JM, Hawkins JM and Hoffbrand HV. (1991) Philadelphia positive acute lymphoblastic leukemia in adults: age distribution, BCR breakpoint and prognostic significance. *Leukemia*, **5**, 196–199.

Stein CA. (1992) Anti-sense oligodeoxynucleotides – promises and pitfalls. *Leukemia*, **6**, 967–974.

Thirman MJ, Gill HJ, Burnett RC *et al.* (1993) Rearrangement of the *MLL* gene in acute lymphoblastic and acute myeloid leukemias with 11q23 chromosomal translocations. *N. Engl. J. Med.* **329**, 909–914.

Tkachuk DC, Kohler S and Cleary ML. (1992) Involvement of a homolog of *Drosophila trithorax* by 11q23 chromosomal translocations in acute leukemias. *Cell*, **71**, 691–700.

Yanuck M, Carbone DP, Pendleton CD, Tsukui T, Winter SF, Minna JD and Berzofsky JA. (1993) A mutant *p53* tumor suppressor protein is a target for peptide-induced $CD8^+$ cytotoxic T-cells. *Cancer Res.* **53**, 3257–3261.

Yokota S, Hansen-Hagge TE, Ludwig W-D, Reiter A, Raghavachar A, Kleihauer E and Bartram CR. (1991) Use of polymerase chain reactions to monitor minimal residual disease in acute lymphoblastic leukemia patients. *Blood*, **77**, 331–339.

Zuelzer WW and Cox DE. (1969) Genetic aspects of leukemia. *Semin. Hematol.* **228**, 228–249.

6

The molecular pathology of neuroendocrine tumours

A.E. Bishop and J.M. Polak
Department of Histochemistry, Royal Postgraduate Medical School, Hammersmith Hospital, London, UK

1. Introduction

During recent years, the field of pathology has switched emphasis from study of the morphological basis of disease to investigation of disease mechanisms at the molecular level. Immunocytochemistry started the advance towards molecular pathology by providing specific means to phenotype cells. The later advent of molecular biology, with the techniques of nucleic acid hybridization and gene amplification, has had a great impact on biological research in general, and its application in pathology is still at a relatively early stage.

Neuroendocrine tumours form a heterogeneous group of neoplasms, united by their possession of secretory granules and potential to release regulatory factors (peptides, amines, etc.). They tend to show benign features on conventional histology (*Figure 1*). These tumours have been variously named over the years and a range of terms for them can still be found in modern literature. Common terms include APUDomas and carcinoid tumours, whilst the more archaic ones of neurocristoma and paracrinoma seem to have been abandoned. It is now the consensus opinion of endocrine pathologists that neoplasms of cells of the diffuse neuroendocrine system should be called neuroendocrine tumours. For the histopathologist, further characterization is based on product identification and signs of malignancy but speculations should not be made

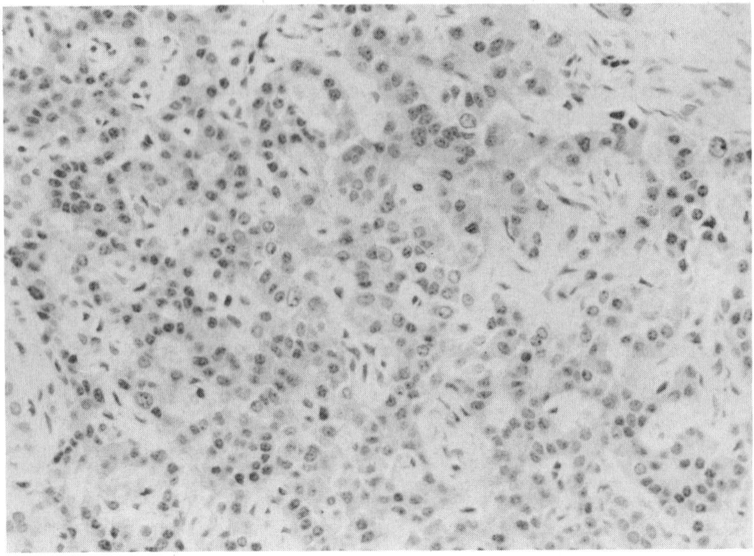

Figure 1. Conventional histology (haematoxylin and eosin stain) of a pancreatic islet cell tumour. Fairly regular cells, with abundant eosinophilic cytoplasm, are arranged mainly in cords separated by fine, fibrovascular stroma. × 170.

on functional activity. Thus, a typical histopathological classification would be, for example, neuroendocrine cell tumour of the pancreas immunoreactive for insulin, rather than a clinical diagnosis of insulinoma.

The use of immunocytochemistry in histopathology, augmented where necessary by ultrastructural study, has rendered neuroendocrine tumours relatively easy to diagnose and classify. However, poor differentiation, production of aberrant molecules, low synthetic activity and the lack of reliable indices for malignancy can hamper diagnosis and prognosis. Therefore, questions about neuroendocrine tumours which remain to be answered include: how is peptide/amine synthesis controlled at the molecular level? Is synthesis of aberrant molecular forms related to abnormal gene expression? Can analysis of gene expression determine malignant potential? The following sections describe the current application of molecular pathology to the study of neuroendocrine tumours and shows how much progress has been made towards answering these and other pertinent questions.

2. Phenotyping: product detection and characterization at the cellular level

2.1. Neuroendocrine cell product detection at the cellular level

The current major means for detection of endocrine cell products is immunocytochemistry, carried out at the light or electron microscopic levels (see Polak and Van Noorden, 1986, and Polak and Priestley, 1992, for review). Requirements for successful immunocytochemical detection include intracellular synthesis of the product to an immunodetectable form, sufficient levels of product storage and the use of highly specific, well characterized immunological probes. Immunocytochemistry, using antibodies to specific (*Table 1; Figure 2*) or general (*Table 2; Figure 3*) markers of neuroendocrine cells, is a routine method in histopathology. One of the main limitations of the phenotyping of neuroendocrine neoplasms with immunocytochemistry is that increased secretory activity of the cells, with attendant depletion of gene products, may often preclude cellular identification, even when the level of gene expression is high. *In situ* hybridization of specific nucleic acid sequences for the detection of gene expression provides not

▶ p. 91

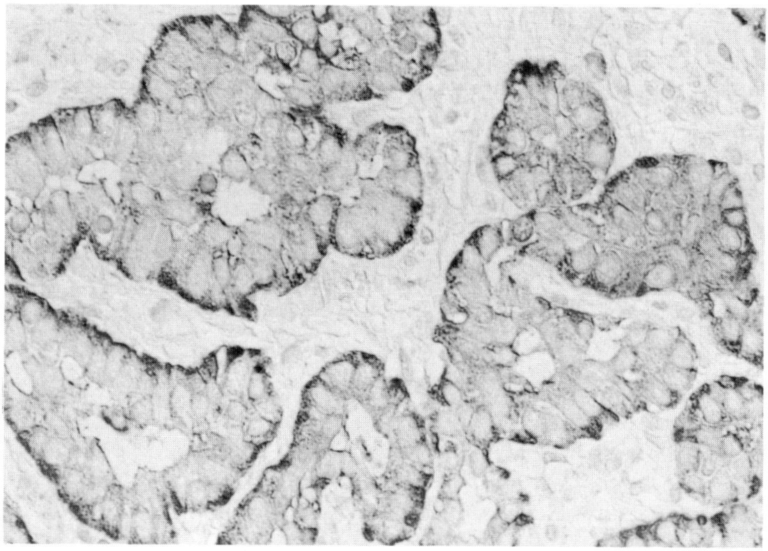

Figure 2. Chromogranin immunoreactivity in nests of cells of an argentaffin (serotonin-producing) carcinoid tumour of the small intestine. (Avidin–biotin–peroxidase (ABC) method with haematoxylin counterstain.) × 255.

Table 1. Neuroendocrine products

Product	Main distribution	Site	Main known actions
Acetylcholine	Central and peripheral nervous system	N	Parasympathetic
Adenosine triphosphate (ATP)	Central and peripheral nervous system	N	Inhibitory neurotransmission
Adrenocorticotrophin (ACTH)	(Anterior) and intermediate lobes of pituitary gland, central nervous system	C, N	Production of adrenal corticosteroids
Amylin	Pancreas	C	Carbohydrate metabolism
Atrial and other natriuretic peptides	Heart	C[a]	Sodium excretion
Bombesin	Amphibian skin and gastrointestinal tract	C	
Calcitonin	Thyroid gland	C	Blood calcium
Calcitonin gene-related peptide (CGRP)	Central and peripheral nervous system	N	Vasodilation; sensory
CCK-8	Central and peripheral nervous system	N	Excitatory
Cholecystokinin (CCK)	Small intestine	C	Gall bladder contraction, pancreatic enzyme secretion
Chorionic gonadotropin	Placenta, anterior pituitary gland	C	Luteotrophic
Corticotrophin-like intermediate lobe peptide (CLIP)	Intermediate lobe of pituitary gland	C	

Corticotropin releasing factor (CRF)	Hypothalamus	N	ACTH production
Dopamine	Central and peripheral nervous system	N	?Sympathetic
Dynorphin	(Anterior) and intermediate lobes of pituitary gland, central nervous system	C, N	Opioid
Endorphin (α and β)	Anterior and intermediate lobes of pituitary gland, central nervous system	C, N	Opioid
Endothelin	Endothelium, central and peripheral nervous system, some endocrine cells	C, N, E	Vasoconstriction
Enkephalin (met- and leu-)	Central and peripheral nervous system, adrenal medulla	C, N C	Opioid
Enteroglucagon	Intestine	C	Trophic to gut
Follicle-stimulating hormone (FSH)	Anterior pituitary gland	C	Ovulation, gonadal hormone production
GABA	Central and peripheral nervous system	N	Modulation of cholinergic activity
Galanin	Central and peripheral nervous system	N	Muscle constriction
Gastric inhibitory polypeptide (glucose-dependent insulinotrophic peptide) (GIP)	Small intestine	C	Insulin secretion, gastric acid secretion
Gastrin	Pyloric stomach and small intestine	C	Gastric acid secretion

Table 1. Neuroendocrine products (cont'd)

Product	Main distribution	Site	Main known actions
Gastrin-releasing peptide (GRP) (see also Neuromedin C and bombesin)	Lung, central and peripheral nervous system	C, N	Trophic, release of other regulatory peptides
GAWK	Central and peripheral nervous system, endocrine system	C, N	
Glucagon	Pancreatic islets (α cells)	C	Blood sugar
Growth hormone	Anterior pituitary gland	C	Growth
Growth hormone-releasing factor (GRF)	Hypothalamus	N	Growth hormone production
Histamine	Peripheral nervous system and gut endocrine system	C, N	Acid secretion
Insulin	Pancreatic islets (β cells)	C	Blood sugar
β-Lipotrophin	Anterior and intermediate lobes of pituitary gland, central nervous system	C, N	Lipolysis
Luteinizing hormone (LH)	Anterior pituitary gland	C	Ovulation, gonadal hormone production
Luteinizing hormone-releasing hormone (LHRH)	Hypothalamus	N	Release of gonadotrophins
Melanocyte-stimulating hormone (MSH) (α, β)	(Anterior) and intermediate lobes of pituitary gland, central nervous system	C, N	Skin pigmentation
Motilin	Small intestine	C	Gut motility
Neural peptide with tyrosine (NPY)	Central and peripheral nervous system	N	Vasoconstriction

Neurokinin A (substance K neuromedin L); neurokinin B (neurokinin K)	Spinal cord (porcine)	N	Tachykinin
Neuromedin B, neuromedin C (-GRP main form)	Spinal cord (porcine)	N	
Neuromedin U	Central and peripheral nervous system	N	Vasoconstriction
Neurotensin	Intestine, adrenal medulla, central nervous system	C, N	Vasodilation, gastric acid secretion
Noradrenaline	Central and peripheral nervous system	N	Sympathetic
Oxytocin	Central nervous system (neurosecretory)	N	Contraction of uterus and let-down of milk
Pancreastatin	Anterior pituitary, endocrine system	C	Inhibition of insulin release
Pancreatic polypeptide	Pancreatic islets	C	Pancreatic enzyme secretion, gall bladder contraction
Parathyroid hormone	Parathyroid gland	C	Blood calcium
Peptide with C- and N-terminal tyrosine (PYY)	Intestinal system	C	Inhibition of acid secretion and motility
Peptide with histamine	Central and peripheral nervous system	N	Muscle relaxation, secretion
Pituitary adenylate cyclase-activating peptide (PACAP)	Central and peripheral nervous system	N	Adenylate cyclase activation
Placental lactogen	Placenta	C	Lactogenic, trophic, lipolytic

2.2. Quantification of peptide secretion at the cellular level

The measurement of circulating and tissue concentrations of neuroendocrine cell products has relied in the past mainly on radioimmunological methods followed by chemical characterization of the product by chromatography. This approach has been a useful adjunct to morphological studies of neuroendocrine pathology. However, new methods have evolved which allow analysis of product secretion at the cellular level and provide the means for detailed investigation of the morpho-functional activity of both normal and neoplastic neuroendocrine cells. A method which allows identification of neuroendocrine cells and measurement of product release *in vitro* is the reverse haemolytic plaque assay (Smith *et al.*, 1988; Yamada, 1990). The technique comprises incubation of cultured neuroendocrine cells with sheep red blood cells conjugated to staphylococcal protein A. Following addition of specific peptide antibodies and complement, haemolysis occurs and the cell secreting the antigen recognized by the antibodies can be identified by its surrounding clear plaque. The size of this plaque is proportional to the amount of peptide secreted, so the method can be used to monitor peptide release (Yamada *et al.*, 1989). In addition, electron microscopy can be carried out on single plaque-forming cells, for further characterization and morpho-functional correlation (Smith *et al.*, 1988). However, although this assay is widely used in many laboratories and has been a valuable tool in endocrinology, it is not without disadvantages. Relatively high concentrations of antibodies are required, non-specific haemolysis can be caused by the complement, passive antigen release can occur from damaged cells and batches of sheep erythrocytes can vary in quality. An alternative method is the cell blot assay (Kendall and Hymer, 1989), which has been used recently to analyse the secretion of peptides from cells of the anterior pituitary (Cimini *et al.*, 1994; *Figure* 4). In this technique, neuroendocrine cells are incubated on a transfer membrane which takes up the secreted peptide. This secretion can then be quantified using Western blots and comparison with standards. The cell blot assay only requires small amounts of antibodies and is relatively quick and simple to use. In addition, double immunostains can be made, allowing investigation of possible co-secretion of different products from the same cell.

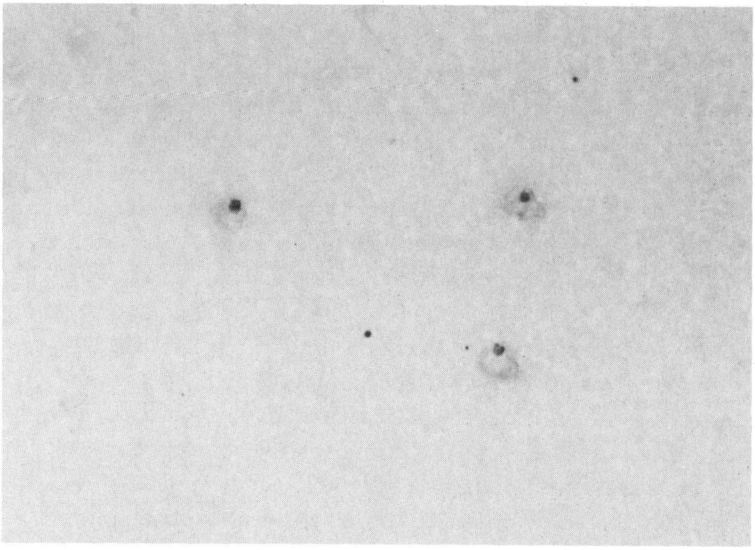

Figure 4. Cell blot of rat anterior pituitary cells stimulated with 0.02 μM adrenocorticotrophic hormone immunostained for vasoactive intestinal peptide (VIP). Haloes can be seen around the cells, signifying release of VIP. × 300.

3. Genotyping: application of molecular probes

3.1. Technology for hybridization analysis

Molecular probes are being used increasingly in endocrine pathology, providing accurate and sensitive detection of gene expression. Two main areas where a molecular approach has been used successfully are *in situ* hybridization, for the morphological localization and characterization of specific nucleic acid sequences, and the cloning and mapping of abnormal genes in inherited endocrine disease (see Section 4.2 on heredity).

Molecular probes can be used in a variety of ways to investigate gene expression in neuroendocrine tumours. These include single filter or dot-blot hybridization, Southern or Northern blotting, polymerase chain reaction (PCR) or *in situ* hybridization. This latter technique allows the localization of particular DNA or RNA species within histological preparations to tissue structures, individual cells or even chromosomes (see Polak and McGee, 1990, for review). It is based on the capacity of complementary nucleic acid sequences to form stable hybrids with generic DNA or RNA. The sequences, or probes, can be prepared in several ways; as single-stranded DNA or RNA, double-stranded DNA or synthetic oligodeoxyribonucleotides. In general, double-stranded DNA probes have been used in the past but single-stranded RNA probes are more sensitive, although they are more difficult to prepare. Synthetic oligonucleotides are becoming more popular and they have several advantages including their total definition of nucleotide sequences, absence of vector and the fact that they can be produced in large quantities.

Visualization of hybridization sites in the tissue preparation depends on efficient direct or indirect labelling of the probes. This can be carried out in a variety of ways. Isotopic labels (e.g. ^{32}P, ^{33}P, ^{35}S, ^{3}H, etc.) incorporated into nucleic acid sequences have been used widely but have a number of drawbacks including safety aspects, poor levels of resolution and delay in visualization. Non-isotopic labels include fluorescence, which has low sensitivity, alkaline phosphatase, peroxidase, bromodeoxyuridine (BrdU), biotin and digoxigenin. Biotinylated probes have been used for some time (Langer *et al.*, 1981) and their sites of hybridization can be visualized with immunocytochemistry. Labelling with digoxigenin is proving very successful, particularly if the detection system is amplified, as it gives high contrast, high resolution demonstration of hybridization (Kennington *et al.*, 1990).

Although *in situ* hybridization is based on the technology of Southern and Northern blotting, its sensitivity, like that of PCR, is far greater; the blotting methods require millions of cells for adequate signal detection whereas *in situ* hybridization can detect a signal of a single gene copy in a chromosomal preparation. The advantage of *in situ* hybridization over PCR has been the capacity to examine intact tissues or cells, allowing direct structural correlations. Although PCR has been used successfully in the past to amplify specific nucleic acid sequences from intact samples of formalin-fixed, wax-embedded tissue, the technology has only recently been developed whereby this amplification can be combined with *in situ* hybridization to give a powerful tool for the investigation of genes with low expression, as often is the case in neuroendocrine tumours (Chen and Fuggle, 1993; Zehbe *et al.*, 1992).

It is possible to quantify the results of *in situ* hybridization, although this has been restricted mainly to analysis of autoradiographs generated by isotopic probes. Silver grains can be counted under a microscope and related to individual cells or tissue components.

3.2. Applications of in situ *hybridization in neuroendocrine tumour pathology*

In situ hybridization is a crucial tool for the endocrine pathologist as it provides the means to look beyond the post-translational events in a cell to examine the site and, possibly, the rate of peptide synthesis, irrespective of the level of peptide storage. Thus, even poorly granulated neuroendocrine tumours can show a high level of gene expression. The small-cell carcinoma of the lung is a prime example of such a tumour, as the cells have scant cytoplasm. This tumour is associated with high circulating levels of chromogranin A, a protein present in secretory granules. In contrast, immunostaining of the tumour for the protein gives no or equivocal results, indicating a high rate of secretion compared with low product storage. *In situ* hybridization of chromogranin mRNA in a series of lung small-cell carcinomas, however, showed strong expression in all cases (Hamid *et al.*, 1991), suggesting that the level of protein synthesis is high. In a similar way, although there is radioimmunological and pharmacological evidence for the production of bombesin (or gastrin-releasing peptide) by these tumours, immunostaining gave equivocal results and it was the application of *in situ* hybridization of probombesin mRNA which showed, at the cellular level, the consistent expression of the gene in these tumours (Hamid *et al.*, 1989).

In situ hybridization yields evidence for hitherto unsuspected production of particular substances by a neuroendocrine tumour. An example of this is carcinoid tumours of the small intestine and lung. Metastatic carcinoid tumours of the small bowel are associated with a well defined clinical syndrome, typically comprising diarrhoea and cutaneous flushing. These tumours are known to arise from the enterochromaffin (EC) cells of the gastrointestinal mucosal epithelium. EC cells produce serotonin (5-hydroxytrypamine), a bioactive amine capable of inducing the features of the clinical syndrome. Pulmonary carcinoid tumours only rarely show production of serotonin. Normal EC cells also produce tachykinins, specifically substance P and neurokinin A, which are derived from β-pre-protachykinin (β-PPT; Nawa *et al.*, 1983). *In situ* hybridization of β-PPT mRNA in a series of ileal and pulmonary carcinoid tumours revealed gene expression in both tumour types, even where serotonin or the tachykinins themselves could not be detected immunocytochemically (Bishop *et al.*, 1990; *Figure 5*).

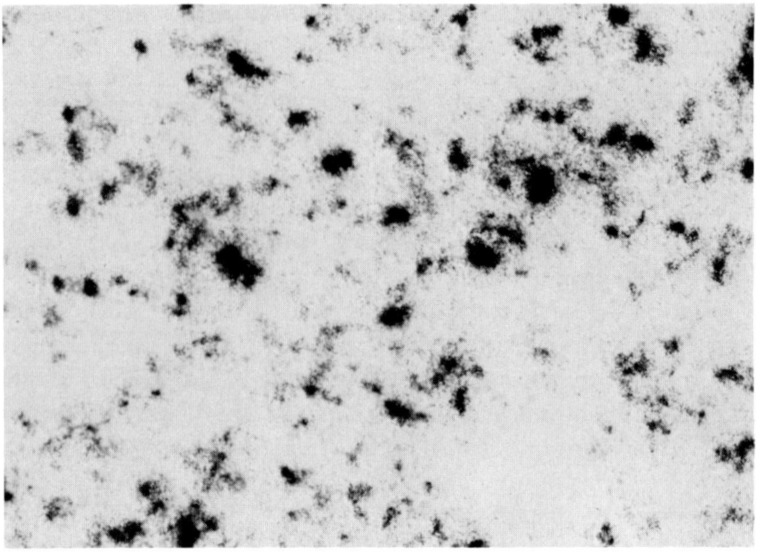

Figure 5. Autoradiograph showing sites of hybridization of β-pre-pro-tachykinin cRNA in a section of a mid-gut carcinoid tumour. × 210.

4. Histogenesis and pathogenesis of neuroendocrine tumours

4.1. Histogenesis

Typically, the morphological appearance of the cells of a neuroendocrine tumour resembles that of its normal counterparts, unlike the heterogeneous patterns seen in non-endocrine tumours. Most neuroendocrine tumours are presumed, or have been proven, to be monoclonal. Some are derived from neuroectodermal stem cells, including medullary carcinomas of the thyroid, paragangliomas, phaeochromocytomas and pituitary adenomas. Others arise from stem cells in the epithelium or tissues originating from the endoderm or mesoderm. This group includes carcinoid tumours and pancreatic neuroendocrine tumours. There are also non-endocrine tumours which show endocrine cell involvement. In the gastrointestinal tract, for example, endocrine differentiation has been noted in gastric (Waldum *et al.*, 1993a,b, and our own observations) and colonic adenocarcinomas (Damjanov *et al.*, 1983; Gibbs, 1967). The endocrine cells are considered to be part of a clonal tumour. The presence of gastrin mRNA has recently been revealed in colonic adenocarcinomas by reverse transcription-PCR (Monges *et al.*,

1993). This finding has particular significance for the biology of these tumours, as gastrin is trophic to cancer cell lines (Sirinek *et al.*, 1985) and gastrin receptors have been found on the cells (Upp *et al.*, 1989).

Mixed endocrine–exocrine tumours of the pancreas are rare, although endocrine cell proliferation can often occur in non-endocrine cancer (e.g. attached to neoplastic glands in ductal adenocarcinomas). However, a true mixed tumour is defined as containing both types of cell in the metastases, in addition to the primary tumour, as well as amphicrine cells, which have both secretory granules and zymogen or mucin vesicles (Klöppel *et al.*, 1993).

4.2. Pathogenesis

Several factors are known to promote carcinogenesis. These include environmental factors, such as chemicals, radiation and viruses, and there is some evidence for their involvement in the initiation and growth of neuroendocrine neoplasms. Genetic factors and the hormonal promotion of tumour growth, while known to be important for a variety of tumour types, have been studied in particular detail in neuroendocrine tumorigenesis.

Heredity. The best characterized examples of the involvement of hereditary factors in the pathogenesis of neuroendocrine tumours are the multiple endocrine neoplasia (MEN) syndromes. In these syndromes, disease inheritance may be dominant or recessive. MEN2 comprises three different, dominantly inherited syndromes: MEN2A is characterized by medullary thyroid carcinoma (MTC) and phaeochromocytoma, MEN2B by MTC with neurofibromatosis and, in familial medullary MTC, the pathology is confined to the thyroid gland. In 1987, evidence for a linked genetic marker and the site of gene deletions were demonstrated in MEN2A (Mathew *et al.*, 1987; Simpson *et al.*, 1987). These occurred on chromosome 10. More recently, specific mutations of the *ret* proto-oncogene on chromosome 10 have been described in MEN2A and familial MTC (Donis-Keller *et al.*, 1993; Mulligan *et al.*, 1993), as well as in MEN2B and sporadic MTC (Hofstra *et al.*, 1994). It has been suggested that the mutations seen in both MEN2A and B cause hyperplasia of cells which normally express *ret*, with neoplastic transformation occurring through mutations in other genes (van Heymingen, 1994). Thus, C cell hyperplasia is always present in inherited MTC.

The situation regarding the molecular basis of MEN1 (hyperplasia or neoplasia of parathyroid, pituitary and endocrine pancreas) is not so well understood. Oncogenesis is thought to involve the unmasking of a recessive mutation of a gene locus on chromosome 11 (11q13), but the gene has yet to be cloned (Larsson *et al.*, 1988).

Hormonal stimulation and receptors. There is good evidence that hormonal stimulation can be a major factor in the growth of not only neuroendocrine but also non-endocrine tumours. It is likely that, for several types of neuroendocrine tumour, initiation (e.g. by a genetic and/or environmental factor) is followed by hormonal stimulation, which promotes tumour growth. Conversely, there may be decreased production of growth-suppressing hormones. In MEN1, for example, hormones produced by a tumour can promote secretion by another endocrine tissue (Asa *et al.*, 1987; Brandi *et al.*, 1986). This sustained stimulation can lead to cellular multiplication and neoplasia. Another example occurs in conditions of sustained hypergastrinaemia, where hyperplasia and eventual neoplasia of EC-like cells of the oxyntic mucosa of the stomach occur (see Polak and Bloom, 1988, for review).

The autocrine control of neuroendocrine tumour growth is a well recognized phenomenon, first described for small-cell carcinomas of the lung. The peptide bombesin, or its mammalian counterpart gastrin-releasing peptide, was found to stimulate the growth of these cancer cells in culture (Cuttitta *et al.*, 1985). The small-cell carcinoma cells produce this factor which acts via cell surface receptors. As the means to convey external influences to endocrine cells, such receptors play a pivotal role in the control of cellular functions. Endocrine cells possess a variety of receptors which can activate or suppress secretion, whose numbers and level of activity can vary. Thus, receptors are likely to be involved in neuroendocrine tumour pathogenesis, although their role has been little investigated as yet. One area which has been studied involves the peptide somatostatin, which is known to inhibit a variety of physiological functions. Clinical studies using the long-acting somatostatin analogue SMS201-995 in the treatment of individuals with pituitary adenomas have shown that it reduces tumour size, decreases the levels of circulating growth hormone and improves the symptoms of acromegaly (Lamberts *et al.*, 1986). As well as being found on human pituitary tumours (Moyse *et al.*, 1985), high-affinity somatostatin receptors have been detected in more than 80% of a series of carcinoid tumours of the gastrointestinal tract and 70–100% of islet cell tumours (Reubi *et al.*, 1991). They are used in tumour therapy, and it is possible to localize neuroendocrine tumours through injection of radiolabelled somatostatin analogue (Janson *et al.*, 1993; Kwekkeboom *et al.*, 1993;

Lamberts *et al.*, 1993). Thus, somatostatin receptors appear to act in the control of endocrine cell secretion and growth but it remains to be seen whether they are involved in tumorigenesis.

Oncogenes. The study of oncogenes in neuroendocrine tumour pathology was initiated relatively recently and conclusive data are scant. There is evidence that c-*myc* amplification enhances growth of small-cell carcinoma cell lines (Carney, 1989; Carney *et al.*, 1985; Takahashi *et al.*, 1989), but such amplification has only been reported rarely in fresh tumours (Takahashi *et al.*, 1989; Wong *et al.*, 1986). Some clinical associations have been reported, however. There is an increased rate of c-*myc* gene activation in relapsing, more aggresive tumours compared with untreated cases (Weynant *et al.*, 1990). Morphological investigations also provide support for a role for c-*myc* in promotion of small-cell carcinoma growth. c-*myc* mRNA expression has been detected in both fresh tumours and cell lines by *in situ* hybridization (Gu *et al.*, 1988) and immunocytochemistry has demonstrated the presence of c-*myc* in 30% of routinely processed tumours (Roncalli *et al.*, 1991). This latter study also found c-*myc* in atypical carcinoids, lending support to the theory that they are more closely related to the aggressive, poorly differentiated lung tumours than to benign carcinoids.

Over-expression of N-*myc* appears to occur in medullary carcinoma of the thyroid. N-*myc* transcript was found in more than 30% of tumours examined (Boultwood *et al.*, 1989) and immunoreactivity for the protein in 20% (Roncalli *et al.*, 1991; *Figure 6*), whilst neither were found in normal C cells. For neuroendocrine tumour of the pancreas, levels of c-*myc* transcript and product appear to be similar or lower than those found in normal endocrine or exocrine pancreas (Hofler *et al.*, 1988; Roncalli *et al.*, 1991). In contrast, over-expression of Ha-*ras* and Ki-*ras* has been detected in pancreatic endocrine tumours, albeit in few cases, and measurement of Ha-*ras* mRNA has been suggested as a means to assess tumour prognosis (Hofler *et al.*, 1988).

Other factors. Some tumours of endocrine glands can be induced by radiation, prime examples being tumours of the thyroid and parathyroid glands (Fleming *et al.*, 1985; Ireland *et al.*, 1985; Katz, 1979; Schneider *et al.*, 1985). Chemical induction of neuroendocrine tumours has often been studied in experimental animals, for example the production of β-cell tumours by streptozotocin and nicotinamide (Yagihashi and Najki, 1981), but no definite association has been demonstrated in humans.

Although viruses have been described in endocrine tumours and are

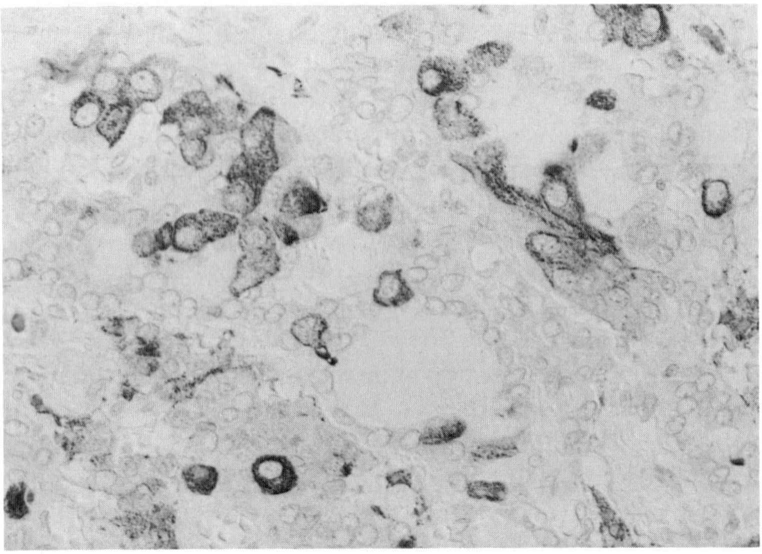

Figure 6. Cells containing dense immunoreactivity for N-*myc* in a medullary carcinoma of the thyroid. (ABC method with haematoxylin counterstain.) × 255.

known to be capable of inducing neoplasia (Kovacs *et al.*, 1976; Temin, 1970), there is no direct evidence for their involvement in neuroendocrine tumorigenesis.

5. Prognostic indicators

Much work has been done to try to establish a histopathological method which can demonstrate reliably the rate of tumour proliferation. A widely used approach is to count mitotic figures, which can be a useful prognostic index in certain cases. However, this method is not without problems; the length of the mitotic phase in tumours can vary and it is often difficult to assess mitosis in inadequately processed tissues (e.g. frozen sections or separated cells). In the past, radioactive methods, especially [^{3}H]thymidine labelling for determining cellular S phase, have been applied in order to determine tumour cell kinetics, but they remain unpopular as they are arduous, time-consuming and involve the use of radioactive substances. A current method for assessing the DNA content of tumours is flow cytometric analysis of cells which have been stained with a substance which binds specifically to nucleic acids. As

DNA content is directly proportional to mitotic activity, the occurrence of abnormal numbers of chromosomes can be detected. Several chemicals have been used to label DNA, including propidium and ethidium, benzamide and acridine orange (Ormerod, 1990). The study of tumour cell kinetics was advanced by the development of monoclonal antibodies which can detect the incorporation of BrdU into intracellular DNA (Gratzner, 1982). BrdU is an analogue of thymidine which can be administered intravenously prior to the removal of tissue biopsies. The labelled cells can then be identified using either flow cytometry of dispersed cells or immunocytochemistry with monoclonal BrdU antibodies on tissue sections (Dolbeare *et al.*, 1983; Hoshino *et al.*, 1986). Another immunocytochemical method for examining tumour proliferation rate is the use of the monoclonal antibodies known as Ki-67, which recognize a nuclear antigen present in proliferating cells. These antibodies were raised against a crude nuclear fraction of a Hodgkin's disease-derived cell line (Brown and Gatter, 1990) and can be used to assess the growth fraction of normal, reactive or neoplastic tissues. Ki-67 (and related) immunostaining has been applied to endocrine pathology; the results of DNA flow cytometry and Ki-67 immunocytochemistry on a series of 16 gastroenteropancreatic tumours were compared. The methods gave similar results and highlighted the heterogeneity of proliferative activity in neuroendocrine tumours, with overlap in proliferative index between benign and malignant tumours (von Herbay *et al.*, 1991). Ki-67 immunoreactivity has been reported to be of uncertain value in thyroid tumour evaluation (Rigaud and Bogometz, 1991). A more recent proliferation marker is the monoclonal antibody Ki-Si (Kreipe *et al.*, 1993), but evidence suggests that it does not match Ki-67 for accuracy. Proliferating cell nuclear antigen (PCNA) has also proved to be useful in determining tumour activity. Following its discovery in 1978 in sera from patients with systemic lupus erythematosus, it was identified as a human auto-antigen present in proliferating but not non-dividing cells (Gerdes *et al.*, 1983). PCNA is present in discrete nuclear regions in S phase, which resemble sites of ongoing cellular DNA synthesis. Antibodies to PCNA may cause overestimation of the proliferating fraction and, although it can be used in paraffin-embedded material, Ki-67 with microwave enhancement may be preferable.

Despite the emergence of these markers for tumour proliferation, many of the problems regarding prognostic indicators for neuroendocrine tumours specifically remain unresolved. DNA aneuploidy in a solid tumour, assessed by flow cytometry, is usually associated with malignant disease and histological evidence of poor differentiation (Atkin and Kaye, 1979). Unfortunately this association does not hold in the assessment of endocrine tumours. For example, growth hormone-secreting adenomas mostly follow a benign clinical course but can exhibit an aneuploid DNA pattern, as assessed by flow cytometry (Hulting *et al.*, 1985, 1989). Similarly, glomus paragangliomas appear to show no correlation between DNA ploidy and clinical course (van der Mey *et al.*, 1990). Another study of pituitary adenomas evaluated the usefulness of DNA index, assessed by flow cytometry, and Ag-NORs (argyrophilic proteins bound to nucleolar organizer regions) as markers of tumour proliferation and found no clear correlation with DNA index and S-fraction Ag-NOR numbers (Stefaneanu *et al.*, 1990). Measurement of DNA content of insulinomas has been reported to be of no prognostic significance (Graeme-Cook *et al.*, 1990a) but, in contrast, a definite relationship was seen in small-cell carcinomas of the lung between DNA content and patient survival rate (Abe *et al.*, 1985). As yet, little is known about BrdU labelling of proliferating endocrine cells but it has been shown to correlate well with mitotic activity in pancreatic B cells (Davidson *et al.*, 1989).

An immunocytochemical marker which has been proposed as a means to assess neuroendocrine malignancy is the presence of α-human chorionic gonadotrophin (α-hCG) in neuroendocrine tumours. This peptide, which has been found in certain non-neuroendocrine tumours such as gonadal germ cell neoplasms, has been reported to be an index of malignancy in pancreatic islet cell tumours (Heitz *et al.*, 1983; Heyderman *et al.*, 1985). However, a more recent study has shown that immunoreactivity for α-hCG does not relate to the prognosis of functioning insulinomas (Graeme-Cook *et al.*, 1990b). In addition to its normal physiological role in the development of the fetus, α-hCG may have a growth-promoting effect in tumours. Thus, the growth rate of lung carcinoma cells *in vitro* is reduced significantly by the addition of α-hCG antibody or induction of α-hCG specific antisense synthesis, by transfection with an expression vector (Rivera *et al.*, 1989).

6. *In vivo* models of neuroendocrine tumours: transgenic mice

During the past decade, investigation of the molecular genetics of cell transformation at the outset of tumour genesis has led to advances in gene transfer technology, amongst which the production of transgenic animals has been particularly useful in the field of neuroendocrine pathology. This transgenic technology involves the microinjection of the relevant DNA into one of the pro-nuclei of a fertilized oocyte. Injected embryos are implanted into pseudopregnant mice and allowed to develop (for reviews, see Hanahan 1988, 1989; Palmiter and Brinster, 1985, 1986). The extent to which the genome of the progeny contains the injected DNA varies and the integration site of the DNA is randomized. Phenotypic differences can occur because of variations in the patterns of transgene expression (Palmiter and Brinster, 1986).

The transgene can be natural or a hybrid construct. For the neuroendocrine system, the early region of simian virus 40 (SV40) has proved to be particularly successful in promoting tumours. This sequence of the oncogene encodes a protein (large transforming antigen or Tag) which can be identified immunocytochemically, a feature which facilitates localization of transforming cells. To produce transgenic models of neuroendocrine tumours, a hybrid oncogene is constructed from the coding region of the oncogene and the regulatory region of a hormone gene. The resulting transgenic animals express the hybrid oncogene in those tissues which normally contain the hormone-regulating sequence. Several examples exist of transgenic models which provide unique means for investigation of the stages of development of endocrine neoplasms. Many have been shown to be useful models of human tumour disease; mice with an SV40–insulin II hybrid gene show pathology which is confined to the endocrine pancreas, where temporal adaptive changes from normal β cells through hyperplasia/dysplasia occur (Hanahan, 1985; Power *et al.*, 1987). The tumour induction, cellular adaptation and growth patterns of the neoplasm showed similarities with human β-cell tumours (Power *et al.*, 1987). In contrast with the site-specific expression which occurs in SV40–insulin II transgenic mice, another hybrid gene formed from the early region of SV40, with a promoter derived from bovine arginine–vasopressin, did not provide transformation of hypothalamic vasopressin cells, as expected, but of pancreatic β-cells (*Figure 7*) and cells of the anterior pituitary (Murphy *et al.*, 1987). This apparent lack of specificity of vasopressin gene expression indicates that it is under the control of several regulatory elements, some of which are not in the hybrid gene. However, the transgenic model provides a useful analogy with human multiple endocrine neoplasia type I syndrome, a familial disorder transmitted as an autosomal semi-dominant gene.

These models also provide insight into the development of the endocrine pancreas. Studies of the ontogeny of the SV40–arg–

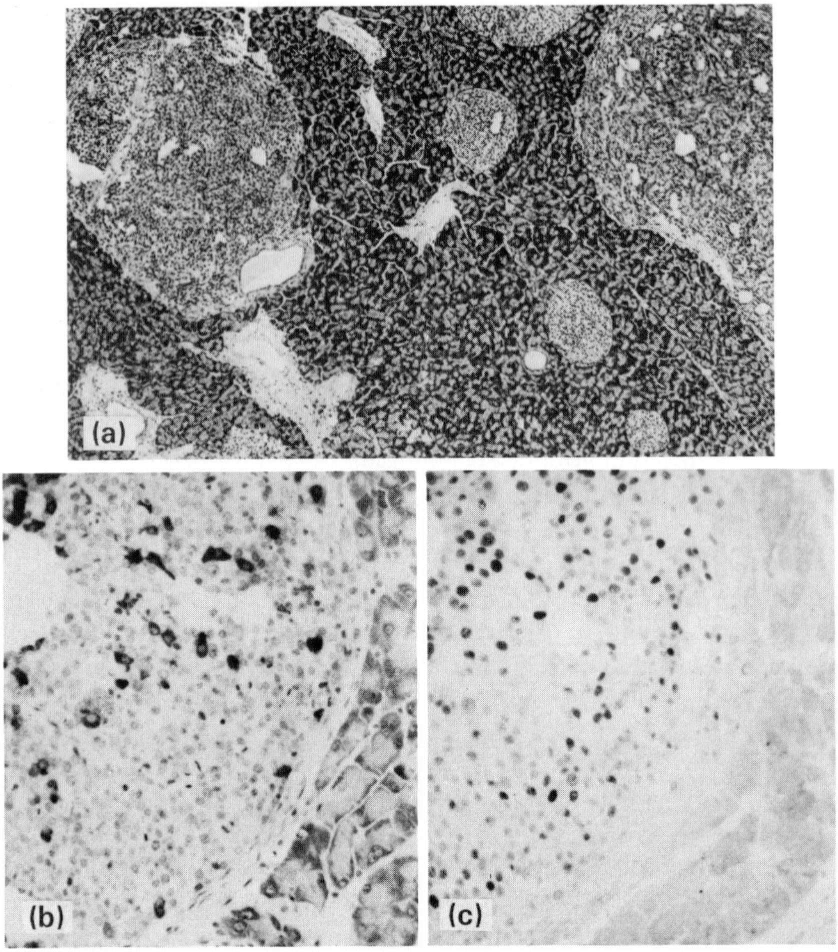

Figure 7. Sections of pancreas from an arginine–vasopressin–SV40 transgenic mouse. (a) Haematoxylin and eosin stain showing normal, hyperplastic and neoplastic islets (× 70); serial sections of a single islet immunostained for pancreatic polypeptide (b) and Tag (c), the latter identifying sites of expression of the hybrid oncogene (peroxidase–anti-peroxidase method with haematoxylin counterstain). × 425.

vasopressin transgenic animals have provided evidence for a common precursor for pancreatic islet cells and an association between these cells and nerves (Asa, 1993). Tag was found first at day 10 of gestation, co-localized with glucagon. Insulin was also found in these cells at day 12 but the Tag and insulin moieties were in cells separate from glucagon by day 14. Transient expression of Tag was also seen in somatostatin and pancreatic polypeptide cells. Animals with a hybrid gene formed from SV40 and glucagon promoter developed A cell tumours or hyperplasia, accompanied by neuronal proliferation in the pancreas (Lee *et al.*, 1992).

7. Conclusions

Major advances in immunology and, more recently, molecular biology have provided the pathologist with new means to investigate the pathobiology of neuroendocrine disease. Neuroendocrine gene products, including structural and functional proteins, specific bioactive products and their corresponding receptor proteins and oncogenes, can be localized with highly refined immunohistochemical procedures. Examination of gene expression, at the level of specific mRNA and genomic DNAs, is being used currently to derive further insight into specific clinical problems as well as allowing new, detailed classification of neuroendocrine tumours.

References

Abe S, Makimura S, Habashi K, Nagai T, Tsuneta Y and Kawakami Y. (1985) Prognostic significance of nuclear DNA content in small cell carcinoma of the lung. *Cancer*, **56**, 2025–2030.

Asa SL. (1993) The endocrine pancreas and its tumours. *Endocr. Pathol.* **4**, 120–130.

Asa SL, Singer W, Kovacs K *et al.* (1987) Pancreatic endocrine tumour producing growth hormone-releasing hormone associated with multiple endocrine neoplasia type I syndrome. *Acta Endocrinol.* **115**, 331–337.

Atkin NB and Kaye R. (1979) Prognostic significance of modal DNA value and other factors in malignant tumours based on 1465 cases. *Br. J. Cancer*, **40**, 210–221.

Bishop AE, Hamid Q, Adams C *et al.* (1990). Expression of tachykinins by ileal and lung carcinoid tumours assessed by combined *in situ* hybridization, immunocytochemistry and radioimmunoassay. *Cancer*, **63**, 1129–1137.

Boultwood J, Wyllie FS, Williams ED and Wynford Thomas D. (1989) N-*myc* expression in neoplasia of human thyroid C cells. *Cancer Res.* **48**, 4073–4077.

Brandi ML, Aurbach GD, FitzPatrick A *et al.* (1986) Parathyroid mitogenic activity in plasma from patients with familial multiple endocrine neoplasia type I. *N. Engl. J. Med.* **314**, 1287–1293.

Brown DC and Gatter KC. (1990) Monoclonal antibody Ki-67; its use in histopathology. *Histopathology*, **17**, 489–503.

Carney DN. (1989) Oncogene and genetic abnormalities in lung cancer. *Chest,* **96** (Suppl.), 25–27.

Carney DN, Gazdar AF, Bepler G *et al.* (1985) Establishments and identification of small lung cancer cells lines having chronic and variant features. *Cancer Res.* **45**, 2914–2923.

Chen RH, Fuggle SV. (1993) *In situ* cDNA polymerase chain reaction. A novel technique for detecting mRNA expression. *Am. J. Pathol.* **143**, 1527–1534.

Cimini V, Van Noorden S, Mahaveda H and Polak JM. (1994). The cell blot assay in analysis of rat anterior pituitary cell secretion. *Histochem. J.* **26**, 59–66.

Cuttitta F, Conway DN, Mulshine J *et al.* (1985) Bombesin-like peptides can function as autocrine growth factors in human small cell lung cancer. *Nature,* **316**, 823–826.

Damjanov I, Amenta PS and Bosman FT. (1983) Undifferentiated carcinomas of the colon containing exocrine neuroendocrine and squamous cells. *Virchows Arch. Pathol. Anat.* **40**, 57–66.

Davidson PM, Campbell IL, Oxbrow L, Hutson JM and Morrison LC. (1989) Pancreatic beta cell proliferation in rabbits demonstrated by bromodeoxyuridine labelling. *Pancreas,* **4**, 594–600.

Dolbeare F, Gratzner H, Pakavinei MG and Gray JW. (1983) Flow cytometric measurement of total DNA content and incorporated bromodeoxyuridine. *Cell Biol.* **50**, 5573–5577.

Donis-Keller H, Don S, Chi D *et al.* (1993) Mutations in the RET proto-oncogene are associated with MEN 2A and FMTC. *Hum. Mol. Genet.* **2**, 881–886.

Fleming ID, Black TL, Thompson EI, Pratt C, Rao B and Hustu O. (1985) Thyroid dysfunction and neoplasia in children receiving neck irradiation for cancer. *Cancer,* **55**, 1190–1194.

Gerdes J, Schwab U, Lemke H and Steion H. (1983). Production of a monoclonal antibody reactive with a human nuclear antigen associated with cell proliferation. *Int. J. Cancer,* **31**, 13–20.

Gibbs NM. (1967) Incidence and significance of argentaffin and Paneth cells in some tumours of the large intestine. *J. Clin. Pathol.* **20**, 826–831.

Graeme-Cook F, Bell DA, Flotte TJ, Pieffer F, Pastel-Levy G, Nardi G and Crompton C. (1990a) Aneuploidy in pancreatic insulinomas does not predict malignancy. *Cancer,* **66**, 2365–2368.

Graeme-Cook F, Nardi G and Compton CC. (1990b) Immunocytochemical staining for human chorionic gonadotrophin sub-units does not predict malignancy in insulinomas. *Am. J. Clin. Pathol.* **93**, 273–276.

Gratzner HG. (1982) Monoclonal antibody to 5-bromo and 5-iododeoxyuridine: a new reagent for detection of DNA replication. *Science,* **218**, 474–475.

Gu J, Linnoila RI, Seibel NL *et al.* (1988) A study of *myc*-related gene expression in small cell lung tumour by *in situ* hybridization. *Am. J. Pathol.* **132**, 13–17.

Hamid QA, Bishop AE, Springall DR *et al.* (1989) Detection of human probombesin mRNA in neuroendocrine (small cell) carcinoma of the lung. *Cancer,* **63**, 266–271.

Hamid Q, Corrin B, Sheppard MN, Huttner WB and Polak JM. (1991) Expression of chromogranin A mRNA in small cell carcinoma of the lung. *J. Pathol.* **163**, 293–297.

Hanahan D. (1985) Heritable formation of pancreatic B cell tumours in transgenic mice expressing recombinant insulin simian virus 40 oncogenes. *Nature,* **315**, 155–122.

Hanahan D. (1988) Dissecting multistep tumourigenesis in transgenic mice. *Annu. Rev. Genet.* **22**, 479–519.

Hanahan D. (1989) Transgenic mice as probes into complex systems. *Science,* **246**, 1265–1275.

Heitz PK, Kasper M, Klöppel G, Polak JM and Vaitukaitis JL. (1983) Glycoprotein hormone alpha chain production by pancreatic endocrine tumours: a specific marker for malignancy. *Cancer,* **51**, 277–282.

von Herbay A, Sieg B, Schurmann G, Hofmann WJ, Betzler M and Otto HF. (1991) Proliferative activity of neuroendocrine tumours of the gastroenteropancreatic system: DNA flow cytometric and immunohistochemical investigations. *Gut,* **32**, 949–953.

Herrington CS, Flannery GMJ and McGee J O'D. (1990) Single and simultaneous nucleic acid detection in archival human biopsies: application of non-isotopic *in situ* hybridization and the polymerase chain reaction to the analysis of human and viral genes. In: In Situ *Hybridization. Principles and Practice* (eds JM Polak and JO'D McGee). Oxford University Press, Oxford, pp. 187–209.

Heyderman E, Chapman DK, Richardson TC, Calvert I and Rosen SW. (1985) Human chorionic gonadotropin and human placental lactogen in extragonadal tumours. *Cancer,* **56**, 2674–2682.

van Heymingen V. (1994) One gene – four syndromes. *Nature,* **367**, 319–320.

Hofler H, Ruhni C, Putz B, Wirnsberger G and Hauser H. (1988) Oncogene expression in endocrine pancreatic tumours. *Virchows Arch. Cell Pathol.* **55**, 355–361.

Hofstra RMW, Landsvater RM, Ceccherini I *et al.* (1994) A mutation in the RET proto-oncogene associated with multiple endocrine neoplasia type 2B and sporadic medullary thyroid carcinoma. *Nature,* **367**, 375–376.

Hoshino T, Nagashima T, Murovic J, Wilson CB, Edwards MSB, Gutin PH, Davis RL and Dearmond SJ. (1986) *In situ* cell kinetics studies on human neuroectodermal tumours with bromodeoxyuridine labelling. *J. Neurosurg.* **64**, 453–459.

Hulting AL, Tribukait B, Bergstrand G and Werner S. (1985) DNA and S-phase representation in human growth hormone-producing pituitary adenomas. *Acta Endocrinol.* **109**, 295–303.

Hulting Al, Askensten U, Tribukait B, Wersalk J, Auer G, Grimelius L, Falkmer S and Werner S. (1989) DNA evaluation in growth hormone-producing pituitary adenomas: flow cytometry versus single cell analysis. *Acta Endocrinol.* **121**, 317–321.

Ireland PJ, Fleming SJ, Levison DA, Cattell WR and Baker LRI. (1985) Parathyroid carcinoma associated with chronic renal failure and previous radiotherapy to the neck. *J. Clin. Pathol.* **38**, 1114–1118.

Janson ET, Westlin JE, Ahlstrom H, Eriksson B, Nilsson S, Akerstrom G and Oberg K. (1993) Somatostatin receptor scintigraphy: recent advances imaging endocrine pancreatic tumours. *Diagn. Oncol.* **3**, 45–48.

Katz AD. (1979) Thyroid and associated polyglandular neoplasms in patients who received head and neck irradiation during childhood. *Head Neck Surg.* **1**, 417–422.

Kendall ME and Hymer WC. (1989) Measurement of hormone secretion from individual cells by cell blot assay. *Methods Enzymol.* **168**, 327–338.

Klöppel G, Höfler H and Heitz PH. (1993) Pancreatic endocrine tumours in man. In: *Diagnostic Histopathology of Neuroendocrine Tumours* (ed. JM Polak). Churchill Livingstone, Edinburgh, pp. 91–122.

Kovacs K, Horvath E and Kerenyi NA. (1976) Intercisternal virus-like particles in a human pituitary adenoma. *Z. Krebsforsch.* **85**, 111–115.

Kreipe H, Heiderbrecht HJ, Hansen S, Rohlk W, Kubbies M, Wacker HH, Tiemann M, Radzun HJ and Parwaresch MR. (1993) A new proliferation associated nuclear antigen detectable in paraffin-embedded tissues by the monoclonal antibody Ki-Si. *Am. J. Pathol.* **142**, 3–9.

Kwekkeboom DJ, Reubi JC, Lamberts SWJ, Bruining HA, Mulder AH, Oei HY and Krenning EP. (1993) *In vivo* somatostatin receptor imaging in medullary thyroid carcinoma. *J. Clin. Endocrinol. Metab.* **76**, 1413–1417.

Lamberts SWJ, Zweens M, Klijn GM, Van Vroonhoven CCJ, Stefanko SZ and Del Pozo E. (1986) The sensitivity of growth hormone and prolactin secretion to the somatostatin analogue SMS 201–995 in patients with prolactinomas and acromegaly. *Clin. Endocrinol.* **25**, 201–212.

Lamberts SWJ, Reubi JC and Krenning EP. (1993) Validation of somatostatin receptor scintigraphy in the localization of neuroendocrine tumours. *Acta Oncol.* **32**, 167–170.

Langer PR, Waldrop AA and Ward DC. (1981) Enzymatic synthesis of biotin-labelled polynucleotides: novel nucleic acid affinity probes. *Proc. Natl Acad. Sci. USA,* **78**, 6633–6637.

Larsson C, Skogseid B, Oberg K, Nakamura Y and Nordenskjöld M. (1988) Multiple endocrine neoplasia type I gene maps to chromosome II and is lost in insulinoma. *Nature,* **332**, 85–87.

Lee YC, Asa SL and Drucker DJ. (1992) Glucagon gene S-flanking sequences direct expression of simian virus 40 large T antigen to the intestine, producing carcinoma of the large bowel in transgenic mice. *J. Biol. Chem.* **267**, 10705–10708.

Mathew CGP, Chin KS, Easton DF *et al.* (1987) A linked genetic marker for multiple endocrine neoplasia type 2A on chromosome 10. *Nature,* **328**, 527–528.

van der Mey AGL, Cornelisse CJ, Hermans J, Terpstra JL, Schmidt PH and Fleuren GJ. (1990) DNA flow cytometry of hereditary and sporadic paragangliomas (glomus tumours). *Br. J. Cancer,* **63**, 298–302.

Monges G, Biagini P, Canstaloube JF *et al.* (1993) Detection of gastrin mRNA in fresh human colonic carcinomas by reverse transcription-polymerase chain reaction. *J. Mol. Endocrinol.* **11**, 223–229.

Moyse E, Le Dafniet M, Epelbaum J *et al.* (1985) Somatostatin receptors in human growth hormone- and prolactin-secreting pituitary adenomas. *J. Clin. Endocrinol. Metab.* **61**, 98–103.

Mulligan LM, Kwok JBJ, Healey CS *et al*. (1993) Germ-line mutations of the RET proto-oncogene in multiple endocrine neoplasia type 2A. *Nature,* **363**, 458–460.

Murphy D, Bishop AE, Rindi G, Murphy MN, Stamp GWH, Hanson J, Polak JM and Hogan B. (1987) Mice transgenic for a vasopressin SV40 hybrid oncogene develop tumours of the endocrine pancreas and anterior pituitary. *Am. J. Pathol.* **129**, 552–566.

Nawa H, Hirose T, Takashima H, Inyama S and Nakanishi S. (1983) Nucleotide sequence of cloned DNAs for two types of bovine brain substance P precursor. *Nature,* **306**, 32–36.

Ormerod MG. (ed.) (1990) *Flow Cytometry. A Practical Approach.* IRL Press, Oxford.

Palmiter RD and Brinster RL. (1985) Transgenic mice. *Cell,* **41**, 343–345.

Palmiter RD and Brinster RL. (1986) Germ line transformation of mice. *Annu. Rev. Genet.* **20**, 465–499.

Polak JM and Bloom SR. (1988) Review: the enterochromaffin-like cell, intragastric acidity and the trophic effect of plasma gastrin. *Aliment. Pharmacol. Ther.* **2**, 291–296.

Polak JM and McGee J O'D. (1990) In Situ *Hybridization. Principles and Practice.* Oxford University Press, Oxford.

Polak JM and Priestley JV. (1992) *Electron Microscopic Immunocytochemistry.* Oxford University Press, Oxford.

Polak JM and Van Noorden S. (1986) *Immunocytochemistry: Modern Methods and Applications,* 2nd edn. John Wright, London.

Power RF, Holm R, Bishop AE, Varndell IM, Alpert S, Hanahan D and Polak JM. (1987) Transgenic mouse model: a new approach for the investigation of endocrine pancreatic β-cell growth. *Gut,* **28**, 121–129.

Reubi JC, Krols L, Krenning E and Lamberts SWJ. (1991) *In vitro* and *in vivo* detection of somatostatin receptors in human malignant disease. *Acta Oncol.* **30**, 463–468.

Rigaud C and Bogometz WV. (1991) Apparent lack of usefulness of monoclonal antibodies Ki-67 in thyroid tumour pathology. *Pathol. Res. Pract.* **187**, 198–200.

Rivera RT, Pasion SG, Wong DTW, Fei YM and Biswas DK. (1989) Loss of tumourigenic potential by human lung tumour cells in the presence of antisense RNA specific to the ectopically synthesized alpha sub-unit of human chorionic gonadotropin. *J. Cell Biol.* **108**, 2423–2434.

Roncalli M, Springall DS, Varndell IM *et al*. (1991) Oncoprotein immunoreactivity in human endocrine tumours. *J. Pathol.* **163**, 117–127.

Schneider AB, Shore-Freedman E, Ryo UY, Bellerman C, Favus M and Pinsky S. (1985) Radiation-induced tumours of the head and neck following childhood irradiation. *Medicine,* **64**, 1–5.

Simpson NE, Kidd KK, Goodfellow P *et al*. (1987) Assignment of multiple endocrine neoplasia type 2A to chromosome 10 by linkage. *Nature,* **328**, 518–530.

Sirinek KR, Levine BA and Roger MP. (1985) Pentagastrin stimulates *in* *vitro* growth of normal and malignant human colon epithelial cells. *Am. J. Surg.* **149**, 35–38.

Smith PF, Lugue EM and Neill JD. (1988) Detection and measurement of secretion from individual neuroendocrine cells using a reverse hemolytic plaque assay. *Methods Enzymol.* **124**, 443–448.

Sporn MB and Roberts AB. (1986) Peptide growth factors and inflammation, tissue repair and cancer. *J. Clin. Invest.* **78**, 329–332.

Stefaneanu L, Murray D, Kovacs K and Smyth HS. (1990) Comparative study of DNA content and interphase nuclear organizer regions in human pituitary adenomas. *Endocrinol. Pathol.* **1**, 177–182.

Takahashi T, Obata Y, Sekido Y *et al.* (1989) Expression and amplification of myc gene family in small cell lung cancer and its relation to biological characteristics. *Cancer Res.* **49**, 2683–2688.

Temin JM. (1970) Malignant transformation of cells by viruses. *Perspec. Biol. Med.* **14**, 11–26.

Upp JR, Singh P, Townsend CM and Thompson JC. (1989) Clinical significance of gastrin receptors in human colon cancers. *Cancer Res.* **49**, 488–492.

Waldum HL, Brenna E, Kleveland PM, Sandvik AK and Syversen U. (1993a) The role of gastric acid inhibitory drugs – physiological and pathophysiological considerations. *Aliment. Pharmacol. Ther.* **7**, 589–596.

Waldum HL, Sandvik AK, Syversen U and Brenna E. (1993b) The enterochromaffin-like cell – physiological and pathophysiological role. *Acta Oncol.* **32**, 141–147.

Weynant P, Humblet Y, Canon JL and Symann M. (1990) Biology of small cell lung cancer: an overview. *Eur. J. Respir. Dis.* **3**, 699–714.

Wong AJ, Rupport JM, Eggleston J, Hamilton JR, Baylin BB and Vogelstein B. (1986) Gene amplification of c-*myc* and N-*myc* in small cell carcinoma of the lung. *Science,* **233**, 461–464.

Yagihashi S and Najki K. (1981) Immunohistochemical and ultrastructural studies on rat islet cell tumours induced by steptozotocin and nicotinamide. *Virchows Arch. Pathol. Anat.* **390**, 181–191.

Yamada S. (1990) The reverse hemolytic plaque assay in endocrine pathology. *Endocrinol. Pathol.* **1**, 129–131.

Yamada S, Asa SL, Kovacs K, Muller P and Smyth HS. (1989) Analysis of hormone secretion by clinically non-functioning human pituitary adenomas using the reverse hemolytic plaque assay. *J. Clin. Endocrinol. Metab.* **68**, 73–80.

Zehbe I, Hacker GW, Sallstrom J, Rylander E and Wilander E. (1992) *In situ* polymerase chain reaction (*in situ* PCR) combined with immunoperoxidase staining and immuno-gold silver staining (IGSS) techniques. *Anticancer Res.* **12**, 2165–2168.

7

Genetic predisposition to breast cancer

M.R. Stratton
Haddow Laboratories, Institute of Cancer Research, Sutton, UK

1. Introduction

Approximately 1 in 10 women in the UK will develop breast cancer during their lifetime. The aetiology of this disease is complex and several risk factors have been identified, including age at onset of menarche and menopause, age at first pregnancy, number of offspring, breast-feeding practice and contraceptive pill use (reviewed in Higginson *et al.*, 1992). It is presumed that many of these are surrogate indices for hormone-mediated susceptibility mechanisms. One unifying hypothesis is that total cumulative exposure of breast epithelium to oestrogen and the associated cumulative mitotic activity (which would, for example, be increased in a woman who has early menarche and late menopause) determines the risk of breast cancer. Environmental factors have also been implicated, mainly through study of migrant populations. For example, in Japan the incidence of breast cancer is low. However, Japanese women who move to the USA acquire the rates of breast cancer of the host population over a couple of generations. Diet has been considered a strong candidate for this environmental effect. However, the precise dietary constituent which is responsible has eluded identification, despite considerable study. In particular, dietary fat intake, which was previously a popular candidate, seems not to be responsible. Finally, study of atomic bomb victims, and of groups subjected to therapeutic or diagnostic radiation, indicates that exposure to ionizing radiation is a risk factor for carcinoma of the breast.

Evidence for genetic susceptibility to breast cancer has been derived from clinical observation, formal epidemiological studies and, most recently, from genetic linkage studies and isolation of the genes responsible. However, breast cancer is a common disease and familial clustering is not always attributable to inherited susceptibility but may also be due to chance association or shared familial risk factors. The interaction of genetic susceptibility with the non-genetic risk factors described above is not at present understood and provides an interesting area of investigation for the future.

2. Clinical studies of familial clustering of breast cancer

Descriptions of familial clustering of breast cancer date back 2000 years to a report in Roman medical literature around AD100. More recently, the 19th century French neurologist, Broca, noticed an unusual number of women with breast cancer in his wife's family. Beyond the primary task of establishing the phenomenon of familial clustering, clinical studies have aimed to categorize breast cancer families into syndromes which are usually characterized by the other classes of cancer in family members and other phenotypic manifestations. Thus, families have been defined as characterized by breast cancer and ovarian cancer; breast cancer, sarcomas, leukaemia, adrenocortical carcinoma and brain tumours; breast cancer and colorectal cancer; and breast cancer only. In Cowden's syndrome, breast cancer is associated with multiple hamartomas of the lips and mucous membranes, adnexal skin tumours and a number of other types of tumour. The impression from these studies is that familial susceptibility to breast cancer is likely to be genetically heterogeneous. The genes concerned all confer an elevated risk of breast cancer (of varying extent) but differ in their other manifestations. Identification of some of the relevant loci and genes has indicated that, overall, this impression is correct and that, even within these carefully defined clinical groups, more than one gene may be responsible.

3. Epidemiological studies of familial breast cancer

Several epidemiological studies have demonstrated an increased risk of breast cancer in the relatives of breast cancer patients compared to the general population (Claus *et al.*, 1990, 1991; Mettlin *et al.*, 1990; Newman *et al.*, 1988). Most show that the risk to relatives increases with decreasing age of onset of the disease in the index case and that the presence of several affected relatives increases the risk more than that of a single affected relative. Segregation analysis has indicated that predisposition to breast cancer is inherited predominantly as a Mendelian autosomal dominant trait which is attributable to a rare gene (or genes) that confers a high risk of the disease. Estimates of the gene frequency of the abnormal allele(s) vary between 0.0006 and 0.008, and the lifetime risk conferred by abnormal alleles has been estimated to be in the order of 80–90%. If there are several genes, however, these estimates will reflect an average of many different penetrances and gene frequencies. Using the largest of these epidemiological studies, the Cancer and Steroid Hormone Study (CASH; Claus *et al.*, 1990, 1991), it has been estimated that 36% cases of breast cancer diagnosed between ages 20 and 29 years are attributable to genetic predisposition. This gradually drops to 1% in cases aged 80 years or more. A large recessive component is excluded by these studies. However, the risk of breast cancer to sisters of affected individuals is higher in many studies than the risk to mothers, a finding which may reflect a minor recessive component or an artefact of ascertainment. It is unclear at present whether there is any residual familial effect which may be attributable to lower penetrance genes.

4. *BRCA1*

Because breast cancer is so common as a sporadic disease and because of the strong possibility of genetic heterogeneity, there was, in the past, considerable scepticism that conventional linkage analysis would be a fruitful approach to the identification of predisposition loci. Indeed, there were a few tentative reports of linkage (which have remained unconfirmed) before the first convincing report of the localization of a breast cancer predisposition gene. Linkage was detected using the marker D17S74 on chromosome 17q in families with early onset breast cancer (Hall *et al.*, 1990). This was confirmed in families with early onset breast cancer and ovarian cancer (Narod *et al.*, 1991). Subsequently, an international consortium that included virtually all groups interested in heritable susceptibility to breast cancer examined 214 families using polymorphic markers on chromosome 17q (Easton *et al.*, 1993, and subsequent papers in *Am. J. Hum. Genet.*, vol. 52, no. 4). This collaborative venture resulted in a substantial narrowing down of the interval in which the gene could reside and a large body of information pertaining to its penetrance and contribution to familial breast cancer. The gene has been formally designated *BRCA1*.

4.1. Mapping of BRCA1

At the time of writing, the cloning of *BRCA1* has not yet been announced. However, it is rumoured to be imminent and by the publication date of this volume it is quite likely that the isolation of the gene will have been reported. To facilitate the cloning process, detailed genetic and physical maps of the region have been constructed (Abel *et al.*, 1993; Anderson *et al.*, 1993; Flejter *et al.*, 1993) and the interval in which the gene is located has been narrowed considerably using recombinants in families, together with information from loss of heterozygosity in sporadic breast and ovarian cancers (see Section 4.4). Because most of this mapping information is not of general interest and will be superceded, it will not be reviewed further here. Nevertheless, it is worth noting that several genes in the region initially recommended themselves as candidates for *BRCA1*. These include *wnt3* (one of the family of genes initially identified in mice as a genomic integration site of mouse mammary tumour virus and, independently, as wingless mutants in *Drosophila*), the *Hox 2* gene cluster (one of the homeobox developmental gene clusters), *Thra-1* (the thyroid hormone receptor gene), *RARA* (the retinoic acid receptor), *NME1* (a gene which appears to be underexpressed in metastases of several different cancer types), *c-erb-B-2* (a growth factor receptor/tyrosine kinase gene that is amplified in many different cancer types, notably breast), *prohibitin* (an inhibitor of cell proliferation which is occasionally mutated in sporadic breast cancers) and *oestradiol 17β-dehydrogenase* (encoding an enzyme involved in steroid metabolism). All have now been excluded, either by mapping or by sequencing the gene in affected members of breast cancer families linked to *BRCA1*. It is therefore likely that *BRCA1* will be a novel sequence.

4.2. The contribution of BRCA1 *to familial breast cancer*

Analysis of pooled information from the international consortium has generated a detailed picture of the clinical syndrome attributable to *BRCA1* and of its contribution to familial breast cancer. Of the 214 breast cancer families included in this study, 57 included at least one case of ovarian cancer, while the remainder were designated as breast cancer only. In families with both breast and ovarian cancer, the best estimate of the proportion of families attributable to this gene was 100% (95% confidence interval (CI) 79–100%) and such a family is illustrated in *Figure 1*. Subsequent analyses, however, have demonstrated that a small minority of families with both breast and ovarian

IARC 2090

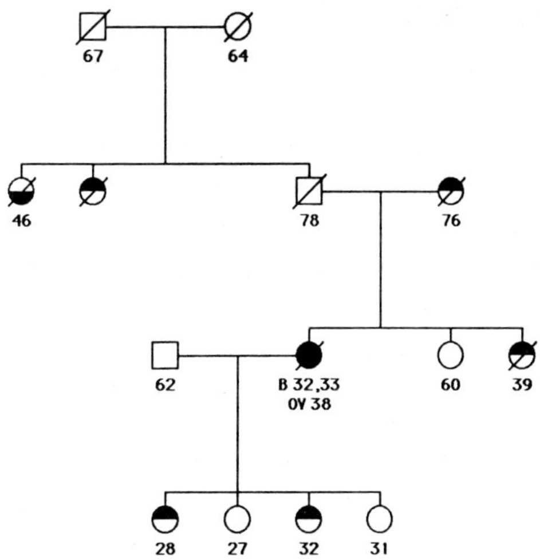

Figure 1. Pedigree of a typical breast–ovarian family that is linked to *BRCA1*. Note the presence of a sporadic ovarian cancer patient marrying into the family. Upper-half shading, breast cancer; lower-half shading, ovarian cancer. The age of diagnosis in affected individuals is indicated. In unaffected individuals the age stated is age of death or present age. Thanks are due to Dr H. Lynch and collaborators for illustration of this family.

cancer are not due to *BRCA1* (Stratton *et al.*, unpublished data, see Section 5.4). Of families with breast cancer only, 45% (95% CI 25–66%) were estimated to be linked to *BRCA1*. Within the latter set, families with a large number of affected cases, and families with very early onset disease, were more likely to be linked to this gene. However, because many were collected for the purpose of locating genes by linkage analysis or because they were so striking, the families included in the consortium analysis are probably subject to considerable ascertainment bias and thus may generate a distorted representation of the contribution of *BRCA1* to breast cancer susceptibility in the general population.

To address this issue further, estimates of the gene frequency of disease-causing alleles of *BRCA1* have been made. By combining penetrance estimates with results of genetic epidemiological studies, it has been possible to derive a preliminary estimate of 0.00066 (95% CI 0.0002–0.001) for the gene frequency of *BRCA1* (Ford *et al.*, 1994b). This would indicate that about 1 in 750 women in the general population carry a *BRCA1* mutation. On this basis, the estimated proportion of breast cancer cases attributable to *BRCA1* would be 17.3% diagnosed below 30 years of age, 7.8% below 40, 3.7% between 40 and 49, and

0.9% in the group of women developing breast cancer at 50 years or over. On these estimates, *BRCA1* probably accounts for about one-third of the familial risk of breast cancer in the general population. Once the gene is cloned and mutations can be determined directly, it should be possible to refine these estimates.

4.3. Risks of cancer in BRCA1 *carriers*

The risks of breast and ovarian cancer in *BRCA1* gene carriers are shown in *Figure 2*. Carriers of a *BRCA1* susceptibility allele have a risk of breast cancer of 50% by age 50 and 85% by age 70 (Easton *et al.*, 1994; Ford *et al.*, 1994). Carriers also have a risk of ovarian cancer of 23% by age 50 and 63% by age 70. The risk of breast cancer rises substantially from age 30, while the risk of ovarian cancer does not appear to increase until age 40. This difference may be of considerable clinical importance in the future, because it will allow most carriers to have their families before considering the prospect of oophorectomy. Whilst the elevated risks of ovarian and breast cancer are immediately apparent from scrutiny of family pedigrees, a more detailed analysis of families linked to *BRCA1* has recently revealed a modest increased risk of colorectal cancer (relative risk (RR) 4.11, 95% CI 2.36–7.15) and of prostate cancer (RR 3.33, 95% CI 1.78–6.20) (Ford *et al.*, 1994). An elevated risk of prostate cancer in relatives of breast cancer cases previously has been apparent in epidemiological studies. It seems unlikely, however, that mutations of *BRCA1* could account for all the excess risk detected by population-based studies.

There is some evidence that variant classes of *BRCA1* allele may exist (Easton *et al.*, 1994). Although major differences in breast cancer risk between *BRCA1* carriers are not apparent, there is evidence that a minority carry an allele that confers a higher risk of ovarian cancer (84% by age 70) than the remainder, in which the risk is predicted to be 32% by age 70 (Easton *et al.*, 1994).

Confirmation of this hypothesis will depend upon characterization of mutations in *BRCA1*, once it is isolated, and will obviously have major implications for clinical management of carriers. There is little evidence at present, however, for independently segregating modifying genes that influence the risk of breast cancer in families attributable to *BRCA1* (Easton *et al.*, 1994).

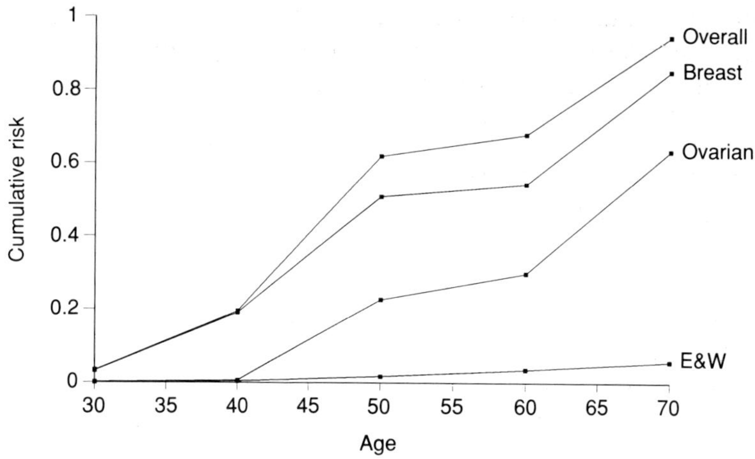

Figure 2. Risks of breast and ovarian cancer in carriers of *BRCA1* mutations. Note that the risk of breast cancer rises before that of ovarian cancer. E&W indicates the age-specific population risks in England and Wales.

4.4. *How do mutations in the* BRCA1 *gene affect the function of its protein?*

The majority of cancer predisposition genes suffer mutations that result in inactivation of the protein that they encode (a notable exception to this rule is the *ret* gene in MEN2A and MEN2B). Often, both alleles have to be mutated in order to inactivate all the cellular protein, and mutation of the second allele is often by loss of heterozygosity (LOH) over a large stretch of DNA that includes the gene. Genes following these patterns of mutational events are often known as tumour suppressor genes or recessive oncogenes, although neither name is completely appropriate.

There is no reason, at present, to predict that *BRCA1* does not conform to the conventional pattern. LOH in the *BRCA1* region is detectable in breast and ovarian cancers arising in *BRCA1*-linked families (Kelsell *et al.*, 1993; Smith *et al.*, 1992). In all tumours analysed so far, the allele lost is the wild-type, i.e. not the one which is linked to cancer susceptibility in the family. This is precisely the pattern expected of a conventional tumour suppressor gene. Moreover, LOH in the vicinity of *BRCA1* is detectable in sporadic breast and ovarian cancers, suggesting that it may be a target for somatic mutation in sporadic tumours, also a conventional pattern of events (Cornelis *et al.*, 1993).

However, there are a number of known or potential tumour suppressor genes on chromosome 17q, including *NF-1* (the neurofibromatosis type 1 gene), *NME, prohibitin* and at least one zone of LOH in sporadic cancers that does not map to the *BRCA1* region (Cornelis *et al.*, 1993; Jacobs *et al.*, 1993). It is, therefore, possible that some or all of the LOH detected in the vicinity of *BRCA1* is aimed at inactivation of genes other than *BRCA1*. If so, it would remain plausible that the *BRCA1* product is being activated by mutations rather than inactivated. Moreover, in published cases of familial tumours it has almost always been the paternal allele that has been lost, a pattern suggestive of a mechanism dependent upon differential genomic imprinting during male and female gametogenesis (Kelsell *et al.*, 1993; Smith *et al.*, 1992). However, transmission of the susceptibility trait through families does not obviously bear the hallmarks of a parental imprinting mechanism and this observation is probably attributable to the structure of the families examined.

5. Other breast cancer predisposition genes

The analysis of the consortium data suggests that over half the families with breast cancer only are not attributable to *BRCA1*. Overall, other genes appear to be less penetrant than *BRCA1* and lead to breast cancer with a later age of onset. Isolation of these genes remains a challenge and is being addressed by consideration of candidate genes and by genomic linkage searches using families unlinked to *BRCA1*. The success of the latter enterprise will depend critically upon the degree of genetic heterogeneity within this set. The contribution of various candidate genes and other known phenotypic features of non-*BRCA1*-linked breast cancer families are reviewed in Sections 5.1–5.4.

5.1. The role of the p53 *gene in familial breast cancer*

p53 was originally identified as a host cell protein bound to a virus-encoded protein (the large T antigen) that is responsible for mediating cellular transformation by the small DNA virus, SV40. It has subsequently been shown that the *p53* gene is mutated as a somatic event in about 50% of human cancers. Following this observation, many groups attempted to identify a cancer predisposition syndrome that was attributable to germline mutations in *p53*, and it transpired that *p53* mutations account for some families with the Li–Fraumeni syndrome (Malkin *et al.*, 1990; Srivastava *et al.*, 1990).

The Li–Fraumeni syndrome was first identified using children with soft tissue sarcomas as index cases in searches of cancer registries for family members that were also affected with sarcomas or other cancers. The syndrome now includes osteosarcomas, gliomas, leukaemias, adrenocortical carcinomas and breast cancers in female relatives (Li *et al.*, 1988). In female carriers, breast cancer is the commonest malignancy of adulthood, and is notable for its particularly early onset. Many women develop breast cancer before the age of 30 years and the large majority of those that develop the disease do so before age 40 (as is the case in the family shown in *Figure 3*). This is in contrast to *BRCA1* carriers, in whom breast cancer before age 30 is rare and in whom many breast cancers develop between 40 and 50 years of age. There are, however, occasional pedigrees attributable to germline *p53* mutations in which the age of onset is later (Prosser *et al.*, 1992) and which may reflect lower penetrance disease-causing alleles of *p53*.

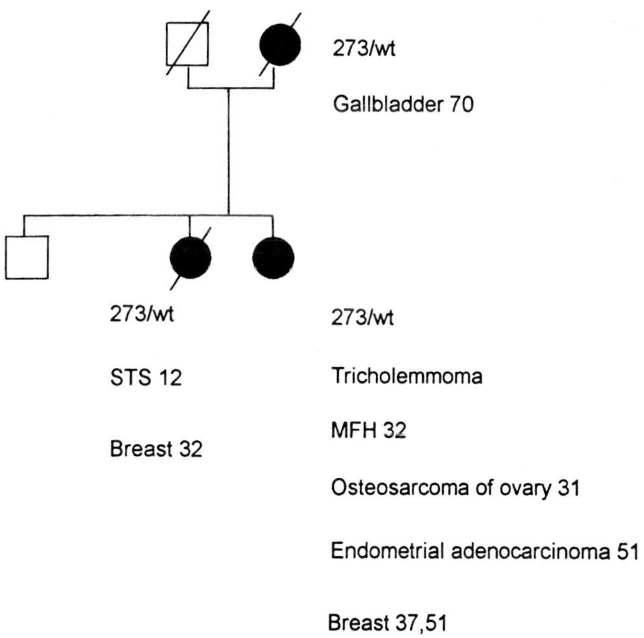

Figure 3. Pedigree of a cancer family due to a germline mutation at codon 273 of the *p53* gene. Note that the mutated allele is being transmitted to the two sisters from their mother who only developed cancer late in life. Whether this apparent lower penetrance in the mother is due to chance, the effect of modifying genes or modifying environmental/lifestyle influences is unknown.

Germline abnormalities of *p53* that result in cancer susceptibility are mainly missense mutations causing amino acid changes within the domains of the protein that are highly conserved through evolution. Frameshift mutations and gross abnormalities of the gene have been reported but are less common. This pattern of germline alterations is similar to the pattern of somatic mutations of *p53* in sporadic cancers, although there remain some notable differences. *p53* mutations are believed to account for at least 50% of families with Li–Fraumeni syndrome, with some families clearly not due to this gene (Santibanez-Koref *et al.*, 1992). Interestingly, there has been one report of a family with breast cancer and sarcomas in which there was intense staining of normal tissues for *p53* protein but no mutation of the gene (Barnes *et al.*, 1992). This pattern of staining is not found in normal tissues of normal people or in the majority of Li–Fraumeni syndrome individuals with germline *p53* mutations, but is a feature of many tumours with somatically acquired *p53* mutations. Therefore it is possible that the cancer in this family is due to germline abnormalities of a different gene, the product of which interacts with p53 protein.

The contribution of germline *p53* mutations to familial breast cancer is small (Prosser *et al.*, 1991; Warren *et al.*, 1992). Even in families unlinked to *BRCA1* there is little evidence from linkage analyses using markers on chromosome 17p of a major role (Stratton *et al.*, unpublished data). Moreover, in consecutive series of early onset breast cancer patients, germline *p53* mutations are only detected in 1% of cases or less (Borreson *et al.*, 1992; Prosser *et al.*, 1992; Sidransky *et al.*, 1992). When a family identified by clustering of breast cancer cases is found to carry a *p53* mutation there are often other characteristic manfestations, such as childhood sarcoma, that give a clue as to the gene responsible.

5.2. The role of the ataxia telangiectasia (AT) gene in susceptibility to breast cancer

AT is a rare autosomal recessive disease characterized by degenerative changes in the brain, particularly affecting the Purkinje cells of the cerebellum, immunodeficiency and predisposition to cancer, especially of the lymphoid system (reviewed in Gatti *et al.*, 1991). AT patients are highly sensitive to the effects of ionizing radiation, as are cultured cells from AT patients. Fusion of cells from AT patients in different families, using radiation sensitivity as an assay, suggests that there are two major and, probably, two minor AT complementation groups. Whether these represent distinct genes is not yet known. However, by genetic linkage analysis of AT families, both of the major complementation groups map to the same region of chromosome 11q23 (Foroud *et al.*, 1991).

A number of epidemiological studies have demonstrated an elevated risk of cancer in AT heterozygotes (Borreson *et al.*, 1990; Pippard *et al.*, 1988; Swift *et al.*, 1991). In particular, the risk of breast cancer in women is raised by a factor of 5–10. The contribution of AT to breast cancer overall is, however, difficult to predict, because of uncertainty surrounding the heterozygote frequency, estimates of which range from 0.007 to 0.07 (Swift *et al.*, 1986). However, if the gene frequency and the penetrance are at the higher ends of the ranges quoted, it has been suggested that AT could account for 10–15% of all breast cancer. Of particular interest is the suggestion of an interaction between exposure to ionizing radiation (usually a low dose delivered for medical purposes) and development of breast cancer in AT heterozygotes.

To test the contribution of AT to familial breast cancer, families that are apparently unlinked to *BRCA1* have been examined using markers flanking AT (Wooster *et al.*, 1993). Overall, no evidence of linkage between breast cancer and AT was detected and, individually, most families were inconsistent by haplotype analysis. Therefore AT must make only a small contribution to familial breast cancer. It is possible to reconcile the findings of epidemiological studies with these results by assuming that the penetrance and prevalence of AT mutations in the population are not high enough to result in much family clustering of breast cancer. Even under such assumptions, however, AT could still make a substantial contribution to apparently sporadic breast cancer. This gene may, therefore, turn out to be an example of a relatively low penetrance but relatively common predisposition gene. Elucidation of this issue will follow cloning of AT and direct comparison of mutation frequencies in population-based series of breast cancer cases and controls.

5.3. The oestrogen receptor in familial late onset breast cancer

Most molecular genetic studies have focussed upon families in which the majority of cases of breast cancer develop before age 50. In principle, however, some familial clustering at a later age could be due to genetic susceptibility rather than just to chance. A few genetic analyses have been performed upon such families and there is one report suggestive of linkage to markers in the vicinity of the oestrogen receptor on chromosome 6q (Zuppan *et al.*, 1991). Given the role of the oestrogen receptor in the development and function of the breast, this gene is an attractive candidate. However, no germline mutations have been published and there is little evidence by linkage analysis or direct mutation screening for mutations in the oestrogen receptor being a common cause of early onset familial breast cancer unlinked to *BRCA1* (Stratton *et al.*, unpublished data).

5.4. Familial male breast cancer and the androgen receptor

Breast cancer in men is approximately 100-fold less common than in women (for review, see Thomas, 1993) with approximately one in 1000 men developing the disease over a lifetime. Established risk factors for this disease include Klinefelter's syndrome, exposure to ionizing radiation and a number of clinical conditions that are likely to be associated with an imbalance of androgens and oestrogens (Thomas *et al.*, 1992).

A family history of both male and female breast cancer is also a risk factor for breast cancer in men (Anderson and Badzioch, 1992; Lenfant-Pejovic *et al.*, 1990; Rosenblatt *et al.*, 1991). Relative risks of 6.1 for men with an affected male relative and of 2.2 for men with an affected female relative have been estimated. The risk to female relatives of an affected male increases with decreasing age of the male case and the risk to men increases with decreasing age of related female cases. In addition, the risk of breast cancer in men increases with increasing numbers of affected first degree relatives. These patterns of risk are similar to those observed in women with female relatives affected by breast cancer (Claus *et al.*, 1990, 1991). The familial clustering indicated by systematic studies is supported by many reports of individual families with multiple cases of male and female breast cancer (Demeter *et al.*, 1990; Everson *et al.*, 1976; Kozak *et al.*, 1986; Marger *et al.*, 1975; Teasdale *et al.*, 1976).

One clue to the molecular basis of predisposition to male breast cancer emerged from study of an unusual family in which breast cancer in two brothers was associated with penoscrotal hypospadias (a developmental abnormality of the penis). Although the latter condition can have many causes, other subtler aspects of the phenotype of the affected individuals suggested an abnormality of androgenization. In a proportion of cases, underandrogenization is caused by abnormalities of the androgen receptor (AR), the protein which mediates all the effects of testosterone and dihydrotestosterone in the development of primary and secondary male sexual characteristics.

Sequencing of the AR gene in this family revealed a mutation of codon 607 in exon 3, resulting in a predicted substitution of glutamine for arginine in the AR protein. This amino acid is located in the second of the two zinc fingers that form the DNA-binding domain of the AR protein (Wooster *et al.*, 1992). The association of AR mutation and male breast cancer has subsequently been confirmed in another family (Lobaccaro *et al.*, 1993a,b). Interestingly, the mutation in the second family alters the amino acid adjacent to the one initially reported, suggesting that abnormalities of this domain of AR may have particular importance with respect to male breast cancer predisposition (*Figure 4*). Formal epidemiological studies are now in progress to assess the risk of breast cancer in individuals with testicular feminization, which is the common clinical presentation of mutations of the AR. However, the possible clustering of mutations in exon 3 in the male breast cancer cases so far reported may explain why there have not been more case reports of breast cancer in testicular feminized males, whose mutations are scattered through the coding sequence of the AR gene.

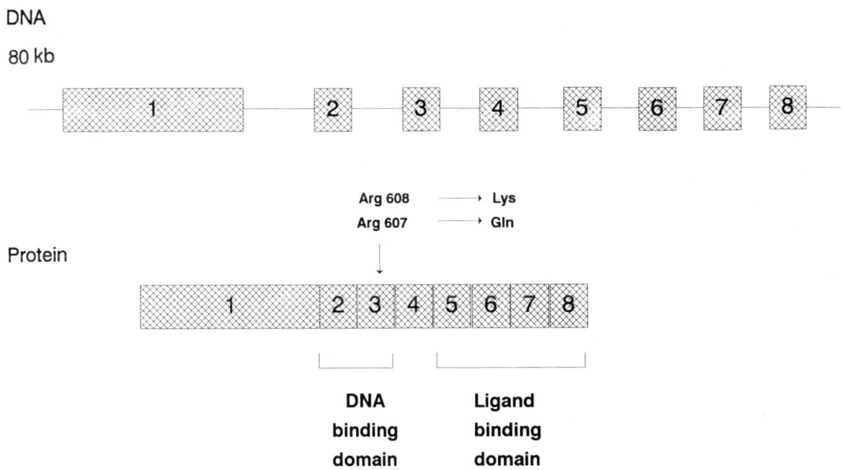

Figure 4. Structure of the androgen receptor gene and protein. Each numbered box represents an exon of the gene. The position within exon 3 of the mutations in male breast cancer cases are indicated. More than 80% missense mutations of the AR in ordinary cases of partial or complete androgen insensitivity are located within exons 5–8 of the gene that encode the ligand-binding domain. Although the amount of information is limited, this tentatively suggests that mutations associated with breast cancer cluster in the second zinc finger of the DNA-binding domain.

Because the AR gene is located on the X chromosome, its role in many pedigrees with male breast cancer can be excluded on the basis of male to male transmission. In addition, by a combination of genetic linkage analysis and direct mutation screening of the AR gene, it appears that it makes only a small contribution to familial male breast cancer (R. Wooster and M.R. Stratton, unpublished data). This conclusion is supported by epidemiological evidence indicating that the risk of breast cancer to second degree paternal relatives is equal to that in maternal relatives of male breast cancer cases (Rosenblatt *et al.*, 1991). Nevertheless, in the light of the difference in breast cancer incidence between men and women, it is of mechanistic interest that abnormalities of genes involved in sexual differentiation can alter breast cancer risk.

The role of *BRCA1* in familial male breast cancer has recently been evaluated. A study of 22 families with at least one case of male breast cancer revealed that only a small minority could be attributable to *BRCA1* (Stratton *et al.*, 1994). Most importantly, it emerged that more than half these families included at least one case of ovarian cancer (e.g. *Figure 5*), and that within this group it is still only a small minority that can be due to *BRCA1*. Since the large majority of ordinary breast–ovarian families are due to *BRCA1*, the gene responsible for familial male breast cancer must confer a much higher risk of male breast cancer than *BRCA1*, despite many similarities in other aspects of the phenotype it generates. Moreover, since the penetrance for male breast cancer in these families is apparently not high, there must be breast cancer and breast–ovarian families lacking male affected cases that are attributable to this gene. Identification of families with male breast cancer will be a useful approach to the identification of families that are unlinked to *BRCA1* in the future and may also enhance the power of linkage searches by reducing genetic heterogeneity.

6. Clinical implications of analyses of breast cancer susceptibility genes

As with other conditions determined by constitutional DNA abnormalities, it is likely that diagnostic testing for susceptibility to breast cancer will become possible, particularly once the genes are cloned. Diagnosis of breast cancer susceptibility will be associated with a number of problems which are medical, ethical and resource-related. Who will be tested? Should all women, or all first degree relatives of affected cases, be offered a test or should it be restricted to those with a

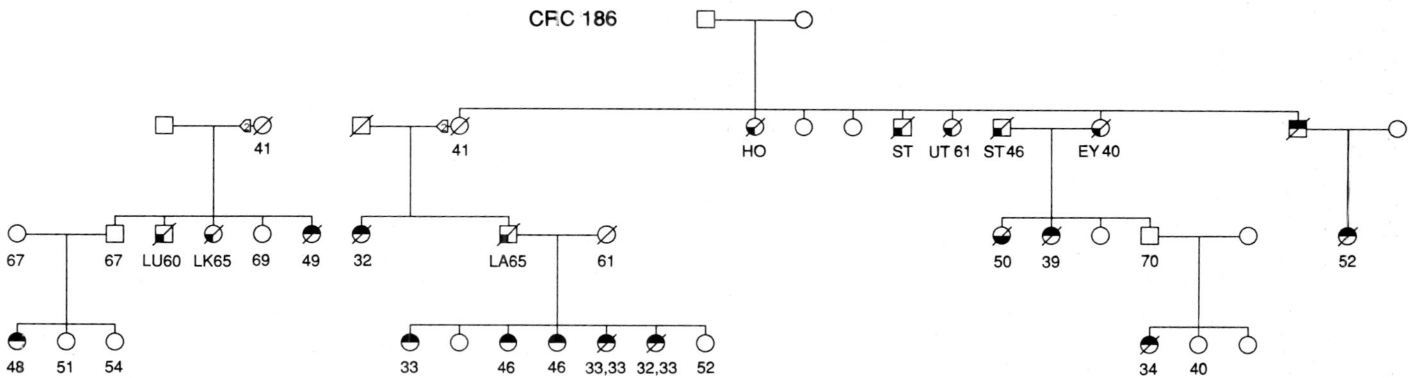

Figure 5. Breast cancer family with a single case of male breast cancer. This family is not linked to *BRCA1*, however the pattern of the remainder of the cancers is similar to that seen in *BRCA1* families. Upper-half shading, breast cancer; lower-half shading, ovarian cancer. The age of diagnosis in affected individuals is indicated. In unaffected individuals the age stated is age of death or present age. Abbreviations for other cancers are: ST, stomach; HO, Hodgkin's' disease; LU, lung; LK, leukaemia; LA, larynx; EY, eye; UT, uterus.

strong family history? Should population testing be offered for genes that may be relatively common but have fairly low penetrance, e.g. AT? Will a positive diagnostic test offer useful avenues to prevention? At present the preventive options range from the relatively acceptable, but probably ineffectual (mammography in young women), to the effective but often unacceptable (prophylactic mastectomy). Should genetic diagnosis be offered for a gene that confers a risk rather than a certainty of a disease – a disease, moreover, that is not always fatal? Will there be a demand for antenatal testing, with subsequent termination of carrier fetuses that would not develop breast cancer for at least 35 years after birth? What are the implications for insurance and employment and what will be the psychological consequences of carrier detection? Many of these issues have, of course, been discussed with respect to other genetic diseases. However, the potential scale of the problem, and of interest in breast cancer, will bring these issues into the public eye considerably more than in the past.

References

Abel K, Boehnke M, Prahalad M *et al.* (1993) A radiation hybrid map of the *BRCA1* region of chromosome 17q12–q21. *Genomics,* **17**, 632–641.

Anderson DE and Badzioch MD. (1992) Breast cancer risks in relatives of male breast cancer patients. *J. Natl Cancer Inst.* **84**, 1114–1117.

Anderson L, Friedman L, Osborne-Lawrence S, Lynch E, Weissenbach J, Bowcock A and King M-C. (1993) High density genetic map of the *BRCA1* region of chromosome 17q12–q21. *Genomics,* **17**, 618–623.

Barnes D, Hanby AM, Gillett CE *et al.* (1992) Abnormal expression of wild type p53 protein in normal cells of a cancer family patient. *Lancet,* **340**, 259–263.

Borreson A-L, Andersen TI, Tretli S, Heiberg A and Meller P. (1990) Breast cancer and other cancers in Norwegian families with ataxia-telangiectasia. *Genes Chrom. Cancer,* **2**, 339–340.

Borreson A-L, Andersen TI, Garber J *et al.* (1992) Screening for *TP53* mutations in breast cancer patients. *Cancer Res.* **52**, 3234–3236.

Claus EB, Risch NJ and Thompson WD. (1990) Age at onset as an indicator of familial risk of breast cancer. *Am. J. Epidemiol.* **131**, 961–972.

Claus EB, Risch NJ and Thompson WD. (1991) Genetic analysis of breast cancer in the Cancer and Steroid Hormone Study. *Am. J. Hum. Genet.* **48**, 232–242.

Cornelis RS, Devilee P, van Vliet M, Kuipers-Dijkshoorn N, Kersenmaeker A, Bardoel A, Meera Khan P and Cornelisse C.J. (1993) Allele loss patterns on chromosome 17q in 109 breast carcinomas indicate at least two distinct target regions. *Oncogene,* **8**, 781–785.

Demeter JG, Waterman NG and Verdi GD. (1990) Familial male breast carcinoma. *Cancer,* **65**, 2342–2343.

Easton DF, Bishop DT, Ford D, Crockford GP and the Breast Cancer Linkage Consortium. (1993) Genetic linkage analysis in familial breast and ovarian cancer: results from 214 families. *Am. J. Hum. Genet.* **52**, 678–701.

Everson RB, Fraumeni JF and Wilson RE. (1976) Familial male breast cancer. *Lancet*, **i**, 9–12.

Flejter W, Barcroft CL, Guo S-W *et al*. (1993) Multicolor FISH mapping with *Alu*–PCR amplified YAC clone DNA determines the order of markers in the *BRCA1* region of chromosome 17q12–q21. *Genomics*, **17**, 624–631.

Ford D, Easton DF, Bishop DT, Narod SA and Goldgar D. (1994) Risks of cancer in *BRCA1* mutation carriers. *Lancet*, in press.

Foroud T, Wei S, Ziv Y *et al*. (1991) Localisation of an ataxia telangiectasia locus to a 3-cM interval on chromosome 11q23: linkage analysis of 111 families by an international consortium. *Am. J. Hum. Genet.* **49**, 1263–1279.

Gatti RA, Boder E, Vinters HV, Sparkes RS, Norman A and Lange K. (1991) Ataxia-telangiectasia: an interdisciplinary approach to pathogenesis. *Medicine*, **70**, 99–117.

Hall JM, Lee MK, Morrow J, Newman B, Anderson l, Huey B and King M-C. (1990) Linkage of early onset familial breast cancer to chromosome 17q21. *Science*, **250**, 1684–1689.

Higginson J, Muir CS and Munoz N. (1992). Breast. In: *Human Cancer: Epidemiology and Environmental Causes.* Cambridge Monographs on Cancer Research. Cambridge University Press, pp. 377–387.

Jacobs IJ, Smith SA, Wiseman RW *et al*. (1993) A deletion unit on chromosome 17q in epithelial ovarian tumors distal to the familial breast ovarian cancer locus. *Cancer Res.* **53**, 1218–1221.

Kelsell DP, Black DM, Bishop DT and Spurr NK. (1993) Genetic analysis of the *BRCA1* region in a large breast/ovarian family: refinement of the minimal region containing *BRCA1*. *Hum. Mol. Genet.* **2**, 1823–1828.

Kozak FK, Hall JG and Baird PA. (1986) Familial breast cancer in males. *Cancer*, **58**, 2736–2739.

Lenfant-Pejovic M-H, Mlika-Cabanne N, Bouchardy C and Auquier A. (1990) Risk factors for male breast cancer: a Franco-Swiss case control study. *Int. J. Cancer*, **45**, 661–665.

Li FP, Fraumeni JF, Mulvihill JJ, Blattner WA, Dreyfus MG, Tucker MA and Miller RW. (1988) A cancer family syndrome in twenty four kindreds. *Cancer Res.* **48**, 5358–5362.

Lobaccaro J-M, Lumbroso S, Belon C *et al*. (1993a) Androgen receptor gene mutation in male breast cancer. *Hum. Mol. Genet.* **2**, 1799–1802.

Lobaccaro J-M, Lumbroso S, Belon C, Galtier-Dereure F, Bringer J, Lesimple T, Heron J-F, Pujol H and Sultan C. (1993b) Male breast cancer and the androgen receptor gene. *Nature Genetics*, **5**, 109–110.

Malkin D, Li FP, Strong LC *et al*. (1990) Germline *p53* mutations in a familial syndrome of breast cancer, sarcomas, and other neoplasms. *Science*, **250**, 1233–1238.

Marger D, Urdaneta N and Fischer JJ. (1975) Breast cancer in brothers. *Cancer,* **36**, 458–461.

Mettlin C, Croghan I, Natarajan N and Lane W. (1990) The association of age and familial risk in a case–control study of breast cancer. *Am. J. Epidemiol.* **131**, 973–983.

Narod SA, Feunteun J, Lynch HT, Watson P, Conway T, Lynch J and Lenoir GM. (1991) Familial breast ovarian cancer locus on chromosome 17q12–23. *Lancet,* **338**, 82–83.

Newman B, Austin MA, Lee M and King M-C. (1988) Inheritance of human breast cancer: evidence for autosomal dominant dominant transmission in high risk families. *Proc. Natl Acad. Sci.* USA, **85**, 3044–3048.

Pippard EC, Hall AJ, Barker DJP and Bridges BA. (1988) Cancer in homozygotes and heterozygotes of ataxia-telangiectasia and xeroderma pigmentosum in Britain. *Cancer Res.* **48**, 2929–2932.

Prosser J, Elder PA, Condie A, MacFayden I, Steel CM and Evans HJ. (1991) Mutations in *p53* do not account for heritable breast cancer: a study in five affected families. *Br. J. Cancer,* **63**, 181–184.

Prosser J, Porter D, Coles C, Condie A, Thompson AM, Chetty U, Steel CM and Evans HJ. (1992) Constitutional *p53* mutation in a non-Li–Fraumeni cancer family. *Br. J. Cancer,* **65**, 527–528.

Rosenblatt KA, Thomas DB, McTiernan A *et al*. (1991) Breast cancer in men: aspects of familial aggregation. *J. Natl Cancer Inst.* **83**, 849–853.

Santibanez-Koref M, Birch JM, Hartley AL, Morris-Jones PH, Craft AW, Eden T, Crowther D, Kelsey AM and Harris M. (1991) *p53* germline mutations in Li–Fraumeni syndrome. *Lancet,* **338**, 1490–1491.

Sidransky D, Tokino T, Helzlsouer K, Zehnbauer B, Rausch G, Shelton B, Prestigiacomo L, Vogelstein B and Davidson N. (1992) Inherited *p53* mutations in breast cancer. *Cancer Res.* **52**, 2984–2986.

Smith SA, Easton DF, Evans DGR and Ponder BAJ. (1992) Allele losses in the region 17q12–21 in familial breast and ovarian cancer involve the wild type chromosome. *Nature Genetics,* **2**, 128–131.

Srivastava S, Zou Z, Pirollo K, Blattner W and Chang EH. (1990) Germline transmission of a mutated *p53* gene in a cancer prone family with Li–Fraumeni syndrome. *Nature,* **348**, 747–749.

Stratton MR, Ford D, Neuhasen S *et al*. (1994) Familial male breast cancer is not linked to the *BRCA1* gene on chromosome 17q. *Nature Genetics,* in press.

Swift M, Morrell D, Cromartie E, Chamberlin AR, Skolnick MH and Bishop DT. (1986) The incidence and gene frequency of ataxia-telangiectasia in the United States. *Am. J. Hum. Genet.* **39**, 573–583.

Swift M, Morrell D, Massey R and Chase CL. (1991) Incidence of cancer in 161 families affected by ataxia telangiectasia. *N. Engl. J. Med.* **325**, 1831–1836.

Teasdale C, Forbes JF and Baum M. (1976) Familial male breast cancer. *Lancet,* **i**, 361–362.

Thomas DB. (1993) Male breast cancer. *Epidemiol. Rev.* **15**, 220–231.

Thomas DB, Jimenez LM, McTiernan A *et al*. (1992) Breast cancer in men: risk factors with hormonal implications. *Am. J. Epidemiol.* **135**, 734–748.

Warren W, Eeles RA, Ponder BAJ ***et al.*** (1992) No evidence for germline mutations in exons 5–9 of the *p53* gene in 25 breast cancer families. *Oncogene,* **7**, 1043–1046.

Wooster R, Mangion J, Eeles R ***et al.*** (1992) A germline mutation in the androgen receptor gene in two brothers with breast cancer and Reifenstein syndrome. *Nature Genetics,* **2**, 132–134 .

Wooster R, Ford D, Mangion J, Ponder BAJ, Easton DF and Stratton MR. (1993) Absence of linkage to the ataxia-telangiectasia locus in familial breast cancer. *Hum. Genet.* **92**, 91–94.

Zuppan P, Hall JM, Lee MK, Ponglikitmongkol M and King M-C. (1991) Possible linkage of the estrogen receptor gene to breast cancer in a family with late onset disease. *Am. J. Hum. Genet.* **48**, 1065–1068.

8

Gene therapy for cancer

M.K.L. Collins
Chester Beatty Laboratories, London, UK

1. Introduction

When the ethics of gene therapy were considered in Britain by the Clothier Committee, it was envisaged that gene therapy would be applied mainly to the treatment of inherited genetic disorders. Indeed, the ethical issues involved in such 'gene replacement therapy' are all that are discussed in the committee's report, published in 1992 (Committee on the Ethics of Gene Therapy, 1992). However, more than half of the subsequent applications for permission to perform gene therapy which have been made to the Clothier Committee, and to its successor, the Lloyd Committee, have involved the treatment of cancer patients. This is, perhaps, due to two influences. Firstly, there is a considerable precedent for the use of experimental treatments for cancer, because often no good alternative exists and the adult patients are capable of giving, and often enthusiastic to give, informed consent. Secondly, there are many technical hurdles remaining before effective, lifetime 'gene replacement therapy' of inherited genetic disorders during childhood will be achieved. In contrast, gene therapy in cancer treatment can be regarded as a short-term, novel method of drug delivery which is technically much less challenging.

2. Cancer gene therapy approaches

2.1. Methods involving gene delivery to tumour cells

Three of the potential routes by which gene therapy may be used to treat cancer are shown in *Figure 1*. Firstly, it may be possible to deliver genes which can revert the transformed phenotype of the tumour cell to

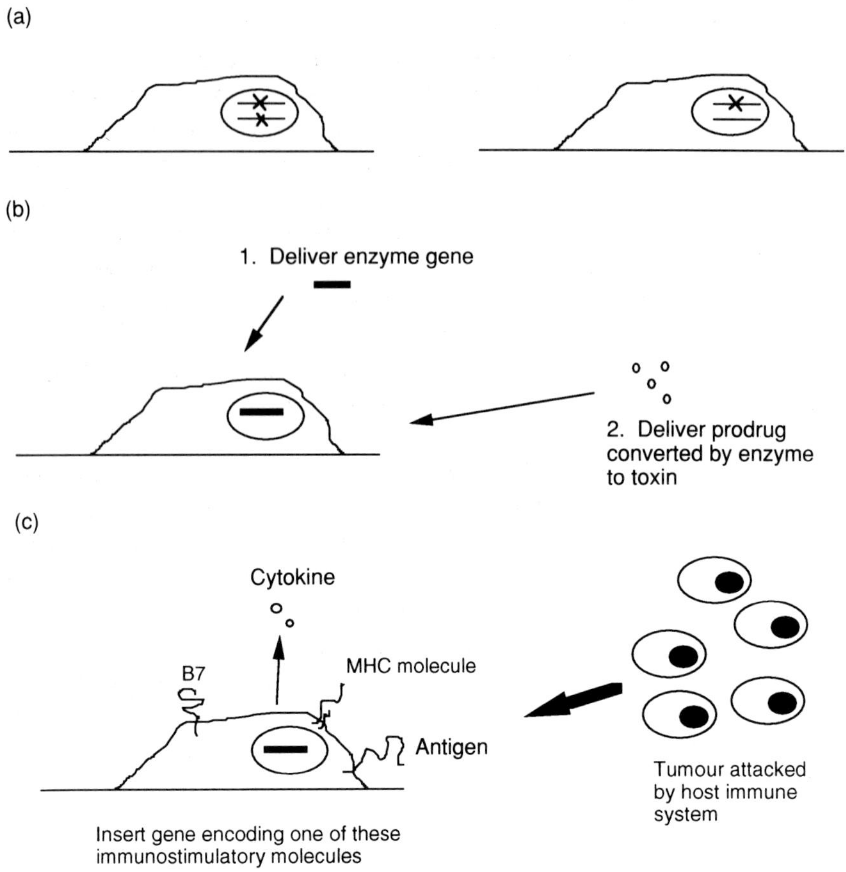

Figure 1. Cancer gene therapy approaches. (a) Repair of genetic defect in tumour cell; (b) enzyme/prodrug delivery; (c) stimulation of host anti-tumour immunity.

that of a more 'normal' cell. This will involve capitalization on the recent advances in understanding of genetic changes in tumour cells which lead to a loss in growth control. For example, many human tumours carry mutations in the recessive oncogene *p53*, and replacement of the *p53* gene in tumour cells has been proposed as a therapy (Friedmann, 1992). Dominantly activating oncogenes could also be inactivated by expressing anti-sense mRNAs, for example to the tumour-specific *bcr–abl* transcript (Skorski *et al.*, 1993), to inhibit their expression, or dominant negative alleles, for example those of the *ras* pathway (Izquierdo *et al.*, 1993), to inhibit their function. Delivering anti-sense mRNA by expression within a tumour cell could be more effective and feasible than the use of anti-sense oligonucleotides.

The second gene therapy approach is to deliver enzymes to tumour cells which are capable of converting inactive prodrugs to cytotoxic derivatives. This principle was first demonstrated using the herpes simplex virus thymidine kinase (HSV-*tk*) gene and the prodrug gancyclovir. When HSV-*tk* was delivered to and expressed in rat brain tumours, the nucleotide analogue gancyclovir was then phosphorylated in the tumour cells. These were killed when it was incorporated into their DNA (Culver *et al.*, 1992). This method has also been used to treat tumour metastases in rat livers (Caruso *et al.*, 1993) and is currently being applied clinically for a variety of tumours (Clinical Trial, 1993a). A number of other enzymes and prodrugs could be used in this way, for example cytosine deaminase can be used to convert 5-fluorocytosine to the cytotoxic 5-fluorouracil (Austin and Huber, 1993). Such enzyme/prodrug systems, which generate cytotoxic nucleotides, require cell division to kill tumour cells. This may not be desirable in some human tumours where the rate of cell division is not rapid.

The third, and currently most popular, group of gene therapy methods are based on the principle of increasing recognition of tumour cells by the host immune system. This relies on the assumption that tumour cells possess antigens which can be recognized as 'foreign' by the host immune effector cells, but that the natural anti-tumour immune response is weak, resulting in tumour escape. Much elegant work on the identification of tumour antigens has been pioneered by Boon and colleagues. They have demonstrated, in a murine mastocytoma, that a rarely expressed, unmutated, cellular protein is the main source of peptide tumour antigen recognized by cytotoxic T lymphocytes, which are the key effector cells in immune control of this tumour (Boon *et al.*, 1980; Uyttenhove *et al.*, 1983). They have extended this work to human melanomas, where they have identified the protein MAGE-1 and tyrosinase, both highly expressed in melanomas, as sources of peptide antigens recognized by patient cytotoxic T lymphocytes (Brichard *et al.*, 1993; Bruggen *et al.*, 1991). It has been proposed that mutated oncogenes might also provide tumour-specific antigens (Melief, 1991). Some B-cell tumours, by virtue of their clonal expression of surface immunoglobulin, carry tumour-specific idiotypes. Finally, many murine tumour antigens were found to be viral in origin, it is therefore possible that antigens from Epstein–Barr virus, human papilloma virus, or human T-cell leukaemia virus might act as tumour antigens in specific human cancers.

Given that tumour antigens do exist, the possibility of boosting the host anti-tumour response (*Figure* 2), by expression of immuno-stimulatory gene products in tumour cells, can be explored. Rejection of animal tumours by their syngeneic hosts has been achieved by expressing, in the tumour cells, molecules which enhance tumour antigen presentation, such as self- or foreign MHC antigens (Hui *et al.*, 1984; Tanaka *et al.*, 1985) and heat-shock proteins (Lukacs *et al.*, 1993), or by expression of viral antigens (Boone and Blackman, 1972; Fearon *et al.*, 1988; Ito *et al.*, 1990; Kobayashi *et al.*, 1969; Sugiura *et al.*, 1988), the co-stimulatory molecule B7 (Townsend and Allison, 1993) and a variety of immuno-stimulatory cytokines (Blankenstein *et al.*, 1991; Colombo *et*

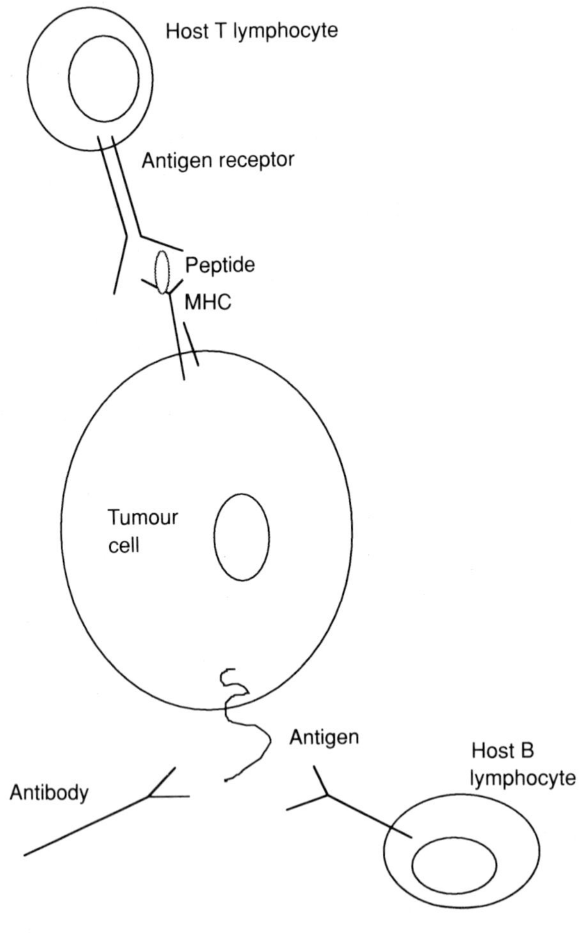

Figure 2. Tumours can look foreign to the immune system.

al., 1991; Dranoff *et al.*, 1993; Esumi *et al.*, 1991; Fearon *et al.*, 1990; Gansbacher *et al.*, 1990a,b; Golumbek *et al.*, 1991; Hock *et al.*, 1991, 1993; Patel *et al.*, 1993; Russell *et al.*, 1991; Tepper *et al.*, 1989; Watanabe *et al.*, 1989). The application of this to patient therapy will require that a strong enough anti-tumour immune response can be generated to attack existing tumour deposits. This has only been demonstrated in animal models for a few of the above genetic modifications. For example, interleukin-2 (IL-2)-secreting tumour cells can reduce the level of metastases when injected into tumour-bearing animals (Caruso *et al.*, 1993; Porgador *et al.*, 1993).

2.2. Indirect cancer gene therapy methods

Section 2.1 described approaches which involved gene transfer to the tumour cells themselves. Clearly, tumour growth involves some co-operation with the host and tumour treatment may have deleterious effects on host cells. Therefore, other cancer gene therapy approaches have been proposed which involve gene transfer to host cells. If solid tumours are to grow beyond a very limited size, they need to be vascularized so that the central cells can receive nutrients. Tumour vasculature is composed of host vascular endothelial cells, which divide rapidly in areas of tumour neovascularization. Conventional therapies which attack tumour vasculature have been shown to be effective, so a gene therapy approach which targeted and eliminated tumour vasculature might be possible.

Dose-limiting toxicity of a number of tumour therapies is reached because host tissues are destroyed. Cells of the bone marrow are particularly sensitive to the radiation and cytotoxic drugs used to kill proliferating tumour cells. It may, therefore, be possible to increase the dose of these treatments by protecting bone marrow cells against their adverse effects. For example, transfer of the DHFR gene to bone marrow stem cells *in vitro*, followed by bone marrow transplant, might allow higher doses of methotrexate to be subsequently used (Corey *et al.*, 1990).

3. Which tumours? Which genes?

3.1. Non-localized disease

The majority of human tumours which are not amenable to existing therapies involve disseminated disease. The only gene therapy approach which can be used for this problem is that of immunostimulation, which can potentially generate a systemic anti-tumour attack. Tumours for which there is best evidence that this may work are those where a basal patient anti-tumour immune response, which is capable of being stimulated, has already been characterized. Melanoma and renal cell carcinoma are examples of such immunogenic tumours which can regress in response to immune stimulation by administration of IL-2 (Rosenberg, 1988). They are therefore popular targets in current and proposed gene therapy trials, most of which involve the injection of tumour cells, genetically engineered to be more immunogenic, into patients with advanced metastatic disease (Clinical Trial 1992a,b,d,e,f, 1993b). It is hoped that information will come from these initial patient trials as to the comparative efficacy of the various strategies, as animal tumour models are, in general, poor predictors of effective immunotherapies. The optimal tumour vaccine may well involve a combination of increased antigenicity, increased antigen presentation and multiple cytokine secretion, following transfer and expression of multiple genes. It has been shown in some patients that B-cell lymphomas will regress in response to the administration of anti-idiotypic antibodies. This is now being attempted by 'genetic immunization'; a vector encoding the patient-specific idiotype is injected into the patient's muscle (Hawkins *et al.*, 1993), a method known to be effective in antibody generation (Ulmer *et al.*, 1993).

3.2. Localized disease

There are a number of tumours which, although they are confined to a single location, are not amenable to surgery. These include primary tumours, such as neural and glial tumours of the central and peripheral nervous system, osteosarcomas and metastatic ovarian tumours contained within the peritoneum. In other cases, multiple metastases may occur within a single organ such as the liver. Direct gene delivery, rather than *ex vivo* modification, can be attempted to treat such tumours. The choice of gene must be dictated by the efficiency of the delivery method and the extent of 'bystander' killing of unmodified cells. With current technology, gene delivery will not approach 100% efficiency and unmodified cells will rapidly escape unless caught in the crossfire. Clearly, immune stimulation can lead to killing of unmodified cells (Plautz *et al.*, 1993) and this is being exploited in a clinical trial in which foreign MHC genes are being directly delivered to tumours (Clinical Trial, 1992c). The gancyclovir treatment of HSV-*tk* modified cells also results in a surprisingly good eradication of unmodified, adjacent cells (Culver *et al.*, 1992). In ongoing trials, the HSV-*tk* gene is being transferred to brain tumour cells *in situ* (Clinical Trial, 1993a). Approaches which propose to revert the transformed phenotype of tumour cells are unlikely to affect the behaviour of unmodified cells, unless secreted paracrine factors are produced. Thus, with current gene transfer technology, they are unlikely to be effective. Indeed, with the problems of assuring access to all cells within a tumour, it is unlikely that gene delivery to tumours will ever be sufficiently sophisticated for these approaches to be possible.

4. Methods of gene delivery

4.1. Physical methods

Direct injection of naked DNA into tumours has been shown to result in gene transfer and expression, with an efficiency that can be of therapeutic benefit (Vile and Hart, 1993b). Liposomes, or DNA–polylysine conjugates, may well prove more suitable for direct gene delivery to tumours and have the additional advantage that they can be targeted using antibodies. Liposomes have been the method of choice for the current clinical trial delivering foreign MHC genes (Clinical Trial, 1992c), and will efficiently deliver genes to a wide variety of tissues *in vivo* (Zhu *et al.*, 1993). A recent innovation in gene delivery has been the demonstration that coupling DNA–polylysine conjugates to adenoviral particles results in escape of the DNA from lysosomes, which leads to much more efficient expression (Curiel *et al.*, 1992), although this has not been shown to be effective in *in vivo* delivery. Physical delivery methods can target a large cell population because a concentrated, large-scale batch of reagent can be prepared. Therefore, they probably remain the most feasible current methods for direct gene delivery to tumours *in situ*.

A major advantage of all physical delivery methods is that there are few size or sequence constraints on the DNA to be delivered. This can allow the use of large genomic fragments of DNA containing multiple control elements. It also avoids incorporation of viral sequences, which can cause safety concerns, or gene expression problems. A further, not inconsiderable advantage of physical methods is that a highly purified preparation of defined composition can be used. All viral gene delivery methods require the production of recombinant viruses from cultured cell lines. The history of contamination of vaccine preparations with adventitious viruses suggests that there is a risk of contamination of such recombinant viral stocks. A problem is that the DNA is only transiently maintained following physical uptake. This may be of less concern in cancer gene therapy than in other gene therapy applications. The delivered gene needs to be expressed only for sufficient time to cause the death of its target cell in tumours, whereas lifetime maintenance of the inserted gene is often desirable for the correction of inherited genetic disorders. It is also possible that this might be overcome by the incorporation of viral sequences which lead to episomal genome maintenance, such as those of EBV or papilloma viruses.

4.2. Viral methods

The first viral vectors to be applied to cancer gene therapy have been recombinant retroviruses. Such viruses have a relatively low titre and are difficult to concentrate. They are therefore not ideal for direct gene delivery to tumours *in situ*. They are being used in several clinical trials to modify tumour cells *ex vivo*, prior to their injection as a vaccine. To be used for *in vivo* delivery, the retroviral titre at the site of the tumour needs to be optimized, for example by the stereotactic injection of retrovirus-producing cells into the tumour. The retroviruses currently being used for human gene therapy applications are based on the murine virus Moloney leukaemia virus, amphotropic strain (MoMLV-A). This is known to be rapidly inactivated on exposure to human serum, because the virus triggers the classical pathway of human complement in an antibody-independent manner (Bartholomew and Esser, 1980). Therefore, it is not suitable for any tumour delivery routes which require prolonged exposure to human serum. MoMLV-A also requires cell division for infection; its genome then integrates into the host cell DNA, so that it is maintained in that cell and its progeny. This is not necessarily a problem in gene delivery to tumour cells, although the rate of cell division in some human tumours, *in vivo* or in culture, is very slow. Recent data suggest that other retroviruses, such as the human immunodefficiency virus type 1, may be able to replicate in non-dividing cells because their core proteins can enter the intact cell nucleus (Bukrinsky *et al.*, 1992). Future developments in recombinant retroviral vector technology may therefore remove the cell division constraint.

The stability of retroviral genome integration and maintenance is an advantage when populations of tumour cells are prepared as vaccines. The modified cells can be cryo-preserved and expanded with little change in the level of foreign gene expression. As retroviruses integrate at random in the genome, there is a potential danger of activation of deleterious genes by insertion of the retroviral enhancer in their proximity. Tumour vaccine preparations are often irradiated, so this problem does not arise. In future, another viral vector with potentially safer mechanism of integration, may be adeno-associated virus (AAV), which integrates into a preferred site on human chromosome 19 in its target cells (Cotin *et al.*, 1992).

The DNA viruses adenovirus (Ad) and herpes simplex virus (HSV) are being developed as vectors for gene therapy applications which require delivery to non-dividing cells. For example, a recombinant Ad has been used in clinical trials for cystic fibrosis, which requires gene delivery to lung epithelium (Kozarsky and Wilson, 1993). HSV vectors are thought to be suitable for gene delivery to neuronal cells because of the ability of the virus to latently infect this target (Andersen *et al.*, 1992). However, it is not clear, as yet, how long gene expression can be maintained *in vivo* after administration of either of these recombinant viruses. They can be prepared as stocks with much higher titres than recombinant retroviruses, which would be a great advantage for direct tumour delivery. Problems which might be encountered in their clinical use include the immunity of most of the population to both viruses and the danger of their recombination with wild-type viruses in the patient.

4.3. Tumour targeting

The technology of clinical gene delivery is really in its infancy, and future cancer gene therapy trials will undoubtedly use more sophisticated vectors and gene delivery methods. It will probably be desirable, particularly when highly toxic therapies are used, to restrict expression of the delivered gene to the tumour cells. This can be achieved by a combination of several levels of tumour targeting (*Figure* 3). Firstly, restricted delivery to the tumour can be achieved by perfusing the tumour in isolation. Secondly, gene uptake can be targeted to the tumour cell surface. This can involve the use of tumour-specific antibodies, either in physical delivery methods, chemically coupled to viral particles (Etienne-Julan *et al.*, 1992) or incorporated into viral surface proteins (Russell *et al.*, 1993). After uptake, gene expression can also be restricted to tumour cells. For example, the murine parvovirus, MVM, replicates only in transformed cells and can be used as a transformed cell-specific delivery vector (Russell *et al.*, 1992). Retroviruses, because of their requirement for cell division, are most likely to

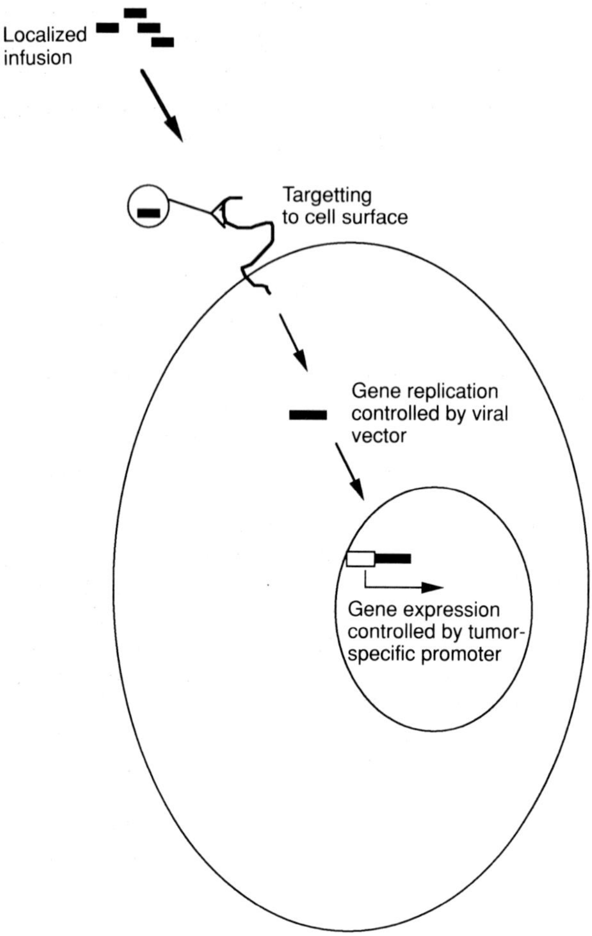

Figure 3. Tumour-targeted gene delivery.

infect tumour cells in the many locations where few normal cells are dividing. Tumour-specific gene regulatory elements may also be incorporated into gene therapy vectors (Vile and Hart, 1993a).

5. Ethical considerations

As many tumours have poor alternative treatments, the risk to the patient of gene therapy can be offset by potential benefits. Thus, gene therapy can be considered in a similar manner to any other experimental tumour therapy. Novel ethical problems only arise when gene therapy poses risks to individuals other than the patient. This could involve transmission of the inserted gene to the germline of the patient, or horizontal transmission of the gene to individuals in contact with the patient. Cancer gene therapy protocols should therefore be reviewed rigorously to minimize the likelihood of these possibilities.

References

Andersen JK, Garber DA, Meaney CA and Breakfield XO. (1992) Gene transfer into mammalian central nervous system using herpes virus vectors. *Hum. Gene Ther.* **3**, 487–499.

Austin EA and Huber BE. (1993) A first step in the development of gene therapy for colorectal cacinoma: cloning, sequencing, and expression of *Escherichia coli* cytosine deaminase. *Mol. Pharmacol.* **43**, 380–387.

Bartholomew RM and Esser AF. (1980) Mechanism of antibody-independent activation of the first component of complement (C1) on retrovirus membranes. *Biochemistry*, **19**, 2847–2853.

Blankenstein T, Qin Z, Uberla K, Muller W, Rosen H, Volk H-D and Diamantstein T. (1991) Tumour suppression after tumor cell-targeted tumor necrosis factor α gene transfer. *J. Exp. Med.* **173**, 1047–1052.

Boon T, Snick JV, Pel AV, Uyttenhove C and Marchand M. (1980) Immunogenic variants obtained by mutagenesis of mouse mastocytoma P815. II. T lymphocyte-mediated cytolysis. *J. Exp. Med.* **152**, 1184–1193.

Boone C and Blackman K. (1972) Augmented immunogenicity of tumour cell homogenates infected with influenza virus. *Cancer Res.* **32**, 1018–1022.

Brichard V, Pel AV, Wolfel T, Plaen ED, Lethe B, Coulie P and Boon T. (1993) The tyrosinase gene encodes an antigen recognised by autologous cytolytic T lymphocytes on HLA-A2 melanomas. *J. Exp. Med.* **178**, 489–495.

van der Bruggen P, Traversari S, Shenez P, Lurquin S, de Plaen E, van der Eynde B, Knuth A and Boon T. (1991) A gene encoding an antigen recognised by cytolytic T lymphocytes on a human melanoma. *Science*, **354**, 1643–1647.

Bukrinsky MI, Sharova N, Dempsey MP, Stanwick TL, Bukrinskaya AG and Haggerty S. (1992) Active nuclear import of human immunodeficiency virus type 1 preintegration complexes. *Proc. Natl Acad. Sci USA*, **89**, 6580–6584.

Caruso M, Panis Y, Gagandeep S, Houssin D, Salzmann JL and Klatzmann D. (1993) Regression of established macroscopic liver metastases after *in situ* transduction of a suicide gene. *Proc. Natl Acad. Sci. USA*, **90**, 7024–7028.

Clinical Trial. (1992a) Immunisation of cancer patients using autologous cancer cells modified by insertion of the gene for interleukin-2. *Hum. Gene Ther.* **3**, 75–90.

Clinical Trial. (1992b) Immunisation of cancer patients using autologous cancer cells modified by insertion of the gene for tumor necrosis factor. *Hum. Gene Ther.* **3**, 57–73.

Clinical Trial. (1992c) Immunotherapy of malignancy by *in vivo* gene transfer into tumors. *Hum. Gene Ther.* **3**, 399–410.

Clinical Trial. (1992d) Phase I study of cytokine-gene modified autologous neuroblastoma cells for treatment of relapsed/refractory neuroblastoma. *Hum. Gene Ther.* **3**, 665–676.

Clinical Trial. (1992e) A pilot study of immunisation with HLA-A2 matched allogeneic melanoma cells that secrete interleukin-2 in patients with metastatic melanoma. *Hum. Gene Ther.* **3**, 677–690.

Clinical Trial. (1992f) A pilot study of immunisation with interleukin-2 secreting allogeneic HLA-A2 matched renal cell carcinoma cells in patients with advanced renal cell carcinoma. *Hum. Gene Ther.* **3**, 691–703.

Clinical Trial. (1993a) Gene therapy for the treatment of brain tumours using intra-tumoral transduction with the thymidine kinase gene and intravenous gancyclovir. *Hum. Gene Ther.* **4**, 39–69.

Clinical Trial. (1993b) Immunisation with interleukin-2 transfected melanoma cells. *Hum. Gene Ther.* **4**, 323–330.

Colombo M, Ferrari G, Stoppacciaro A, Parenza M, Rodolfo M, Mavilio F and Parmiani G. (1991) Granulocyte colony-stimulating factor gene transfer suppresses tumorigenicity of a murine adenocarcinoma *in vivo*. *J. Exp. Med.* **173**, 889–897.

Committee on the Ethics of Gene Therapy. (1992) *Report of the Committee on the Ethics of Gene Therapy*. HMSO, London.

Corey CA, DeSilva AD, Holland CA and Williams DA. (1990) Serial transplantation of methotrexate-resistant bone marrow. *Blood,* **75**, 337–343.

Cotin RM, Linden RM and Berns KI. (1992) Characterisation of a preferred site on human chromosome 19q for integration of AAV DNA by non-homologous recombination. *EMBO J.* **11**, 5071–5078.

Culver KW, Ram Z, Wallbridge S, Ishii H, Oldfield EH and Blaese RM. (1992) *In vivo* gene transfer with retroviral vector-producer cells for treatment of experimental brain tumors. *Science,* **256**, 1550–1552.

Curiel DT, Wagner E, Cotten M, Birnstiel ML, Agarwal S, Li CM, Loechel S and Hu PC. (1992) High-efficiency gene transfer mediated by adenovirus coupled to DNA–polylysine complexes. *Hum. Gene Ther.* **3**, 147–154.

Dranoff G, Jaffee E, Lazenby A, Golumbek P, Levitsky H, Brose K, Jackson V, Hamada H, Pardoll D and Mulligan RC. (1993) Vaccination with irradiated tumor cells engineered to secrete murine GMCSF stimulates potent, specific, and long-lasting anti-tumor immunity. *Proc. Natl Acad. Sci. USA,* **90**, 3539–3543.

Esumi N, Hunt B, Itaya T and Frost P. (1991) Reduced tumourigenicity of murine tumour cells secreting γ-interferon is due to nonspecific host responses and is unrelated to class I major histocompatibility complex expression. *Cancer Res.* **51**, 1185–1189.

Etienne-Julan M, Roux P, Carillo S, Jeanteur P and Piechaczyk M. (1992) The efficiency of cell targetting by recombinant retroviruses depends on the nature of the receptor and the composition of the artificial cell–virus linker. *J. Gen. Virol.* **73**, 3251–3255.

Fearon E, Itaya T, Hunt B, Vogelstein B and Frost P. (1988) Induction in a murine tumor of immunogenic tumor variants by transfection with a foreign gene. *Cancer Res.* **48**, 2975–2980.

Fearon E, Pardoll D, Itaya T, Golumbek P, Karasuyama H, Vogelstein B and Frost P. (1990) Interleukin-2 production by tumour cells bypasses T helper function in the generation of an antitumor response. *Cell*, **60**, 397–403.

Friedmann T. (1992) Gene therapy of cancer through restoration of tumor-suppressor functions? *Cancer*, **70**, 1810–1817.

Gansbacher B, Bannerji R, Daniels B, Zier K, Cronin K and Gilboa E. (1990a) Retroviral vector-mediated γ-interferon gene transfer into tumor cells generates potent and long lasting antitumor immunity. *Cancer Res.* **50**, 7820–7825.

Gansbacher B, Zier K, Daniels B, Cronin K, Bannerji R and Gilboa E. (1990b) Interleukin 2 gene transfer into tumour cells abrogates tumorigenicity and induces protective immunity. *J. Exp. Med.* **172**, 1217–1224.

Golumbek P, Lazenby A, Levitsky H, Jaffee L, Karasuyama H, Baker M and Pardoll D. (1991) Treatment of established renal cancer by tumor cells engineered to secrete interleukin-4. *Science*, **254**, 713–717.

Hawkins RE, Winter G, Hamblin TJ, Stephenson FK and Russell SJ. (1993) Genetic approach to idiotypic vaccination. *J. Immunother.* **14**, 273–278.

Hock H, Dorsch M, Diamantstein T and Blankenstein T. (1991) Interleukin 7 induces $CD4^+$ T cell-dependent tumor rejection. *J. Exp. Med.* **174**, 1291–1298.

Hock H, Dorsch M, Kunzendorf U, Qin Z, Dimantstein T and Blankenstein T. (1993) Mechanisms of rejection induced by tumour cell-targeted transfer of IL2, IL4, IL7, TNF, or IFN-γ. *Proc. Natl Acad. Sci. USA*, **90**, 2774–2778.

Hui K, Grosveld F and Festenstein H. (1984) Rejection of transplantable AKR leukaemia cells following MHC DNA-mediated cell transformation. *Nature*, **311**, 750–752.

Ito T, Wang D-Q, Maru M, Nakajima K, Kato S, Kurimura T and Wakamiya N. (1990) Antitumor efficacy of vaccinia virus-modified tumor cell vaccine. *Cancer Res.* **50**, 6915–6918.

Izquierdo M, Leevers SJ, Marshall CJ and Cantrell D. (1993) p21ras couples the T cell antigen receptor to extracellular signal-regulated kinase 2 in T lymphocytes. *J. Exp. Med.* **178**, 1199–1208.

Kobayashi H, Sendo F, Shirai T, Kaji H, Kodama T and Saito H. (1969) Modification in growth of transplantable rat tumours exposed to Friend virus. *J. Natl Cancer Inst.* **42**, 413–419.

Kozarsky KF and Wilson JM. (1993) Gene therapy: adenovirus vectors. *Curr. Opin. Genet. Devel.* **3**, 499–503.

Lukacs KV, Lowrie DB, Stokes RW and Colston MJ. (1993) Tumor cells transfected with a bacterial heat-shock gene lose tumorigenicity and induce protection against tumors. *J. Exp. Med.* **178**, 343–348.

Melief CJM. (1991) *Adv. Cancer Res.* **58**, 143.

Patel P, Flemming C, Russell S, McKay I, MacLennan K, Box G and Eccles S. (1993) Comparison of the potential therapeutic effects of IL2 or IL4 secretion by a single tumour. *Br. J. Cancer*, **68**, 295–302.

Plautz GE, Yang ZY, Wu BY, Gao X, Huang L and Nabel GJ. (1993) Immunotherapy of malignancy by *in vivo* gene transfer into tumors. *Proc. Natl Acad. Sci. USA*, **90**, 4645–4648.

Porgador A, Gansbacher B, Bannerji R, Tzehoval E, Gilboa E, Feldman M and Eisenbach L. (1993) Anti-metastatic vaccination of tumour-bearing mice with IL2 gene-inserted tumour cells. *Int. J. Cancer*, **53**, 471–477.

Rosenberg S. (1988) Use of tumour infiltrating lymphocytes and interleukin 2 in the immunotherapy of patients with metastatic melanoma. *N. Engl. J. Med.* **310**, 1875–1890.

Russell S, Eccles S, Flemming C, Johnson C and Collins M. (1991) Decreased tumorigenicity of a transplantable rat sarcoma following transfer and expression of an IL-2 cDNA. *Int. J. Cancer*, **47**, 244–251.

Russell SJ, Brandenburger A, Flemming CL, Collins MKL and Rommelaere J. (1992) Transformation-dependent expression of interleukin genes delivered by a recombinant parvovirus. *J. Virol.* **66**, 2821–2828.

Russell SJ, Hawkins RE and Winter G. (1993) Retroviral vectors displaying functional antibody fragments. *Nucleic Acids Res.* **21**, 1081–1085.

Skorski T, Skorska-Nieborowska M, Barletta C, Malaguarnera L, Szczyzlik C, Chen ST, Lange B and Calabretta B. (1993) Highly efficient elimination of Philadelphia leukemic cells by exposure to bcr/abl antisense oligonucleotides combined with mafosfamide. *J. Clin. Invest.* **92**, 194–202.

Sugiura C, Itaya T, Kondoh N, Oikawa T, Kuzumaki N, Takeichi N, Hosokawa M and Kobayashi H. (1988) Xenogenization of tumor cells by transfection with plasmid containing *env* gene of Friend leukaemia virus. *Jap. J. Cancer Res.* **79**, 1259–1263.

Tanaka K, Isselbacher K, Khoury G and Jay G. (1985) Reversal of oncogenesis by the expression of a major histocompatibility complex class I gene. *Science*, **228**, 26–30.

Tepper R, Pattengale P and Leder P. (1989) Murine interleukin-4 displays potent anti-tumor activity *in vivo*. *Cell*, **57**, 503–512.

Townsend SE and Allison JP. (1993) Tumour rejection after direct costimulation of CD8+ T cells by B7-transfected melanoma cells. *Science*, **259**, 368–370.

Ulmer JB, Donnelly JJ, Parker SE *et al.* (1993) Heterologous protection against influenza by injection of DNA encoding a viral protein. *Science*, **259**, 1745–1749.

Uyttenhove C, Maryanski J and Boon T. (1983) Escape of mouse mastocytoma P815 after nearly complete rejection is due to antigen-loss variants rather than immunosuppression. *J. Exp. Med.* **157**, 1040–1052.

Vile RG and Hart IR. (1993a) *In vitro* and *in vivo* targeting of gene expression to melanoma cells. *Cancer Res.* **53**, 962–967.

Vile RG and Hart IR. (1993b) Use of tissue-specific expression of the herpes simplex virus thymidine kinase gene to inhibit growth of established murine melanomas following direct intratumoral injection of DNA. *Cancer Res.* **53**, 3860–3864.

Watanabe Y, Kuribayashi K, Miyatake S, Nishihara K, Nakayama E-I, Taniyama T and Sakata T-A. (1989) Exogenous expression of mouse interferon γ-cDNA in mouse neuroblastoma C1300 cells results in reduced tumorigenicity by augmented anti-tumor immunity. *Proc. Natl Acad. Sci. USA*, **86**, 9456–9460.

Zhu N, Liggitt D, Liu Y and Debs R. (1993) Systemic gene expression after intravenous DNA delivery into adult mice. *Science*, **261**, 209–211.

9

Retrovirus receptors on human cells

R.A. Weiss
Chester Beatty Laboratories, Institute of Cancer Research, London, UK

1. Introduction

Today everyone thinks of retroviruses as giving us AIDS. This is true, of course, although there are human retroviruses (HTLV-I and -II) that also cause leukaemia and lymphomas. Both HTLV-I and the AIDS viruses, HIV-1 and HIV-2, also cause neurological disease, so there is a broad group of diseases to study for which human retroviruses are aetiological agents. But, to give retroviruses their due, they have been kind to medical researchers too, in that they first revealed oncogenes to us and they have given us reverse transcriptase so that we can make cDNA. Moreover, they are providing useful vectors for gene therapy, as described elsewhere in this book.

AIDS is a minor disease in the UK; it has caused about 5000 deaths in the 10 years since we had our first AIDS case. On the other hand, in the USA, AIDS has become the leading cause of death in adult males under 45 years of age. If one considers the years of potential life lost, rather than the illnesses of old age, AIDS is already a leading cause of death, and it will soon surpass cancer and heart disease. In sub-Saharan Africa the incidence of the disease is much worse. Moreover, the WHO estimates that within 10 years there may be more AIDS cases in India than in Africa. It has also become a serious problem in other countries like Thailand and Brazil. Looking to the future, HIV is quite likely to overtake malaria as the major cause of death caused by an infectious microorganism. Demographically, its impact is difficult to exaggerate and, as yet, there is no cure or preventive vaccine in sight.

This chapter will discuss how retroviruses get into cells. *Figure 1* shows a simplified replication cycle of the retrovirus. Like all viruses, a retrovirus is an intracellular parasite with a particular life-cycle. In order to initiate infection, it has to interact with receptor molecules on the cell surface and then become internalized. Much is now known about initial interactions, the binding to cell surface receptors, especially the CD4 receptor for HIV, but rather little is understood, as yet, about the process of internalization. There are various assays which can be used to probe virus–receptor interactions. First, there are binding assays for the whole virus particle or the outer envelope glycoproteins, and one can screen whether antibodies to cellular receptors or candidate receptors block binding. Secondly, there are fusion assays, based either on fluorescence dequenching, when a heavily labelled virus envelope is fused with the cell membrane, or cell-to-cell fusion between infected

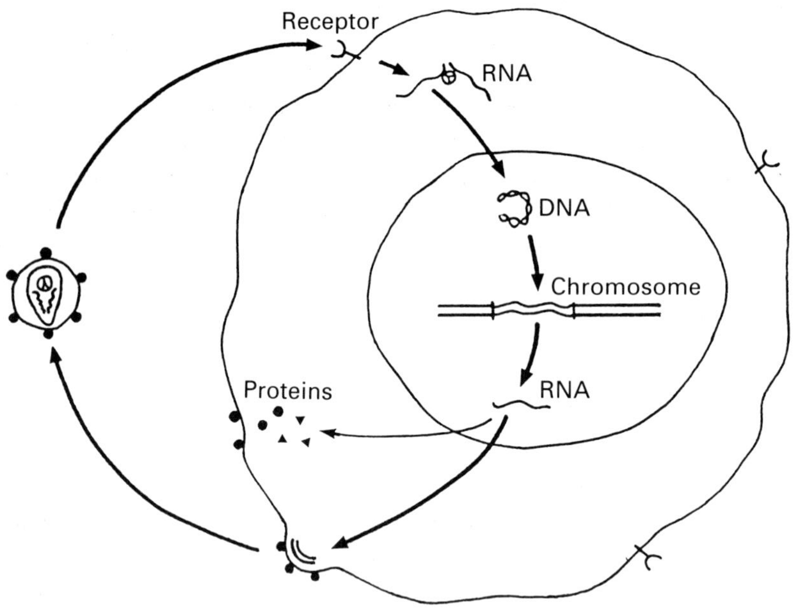

Figure 1. Simplified replication cycle of a retrovirus. The virus particle containing two RNA genomes binds to cell surface receptors and becomes internalized. Reverse transcriptase within the core of the virus copies the RNA into a single DNA provirus which integrates into the chromosomal DNA of the cell. The proviral genome may remain latent, replicating only as part of the cell genome. If its RNA is expressed, however, viral proteins are synthesized and full-length RNA transcripts are packaged in the progeny virus particles.

cells that produce the retrovirus and those that express receptors. With many retroviruses, the interactions of the glycoprotein of the retrovirus on one cell and the receptors on the other lead to cell-to-cell fusion to form large multinucleated syncytia. Cell fusion is a rather crude assay but it is quite specific; by this method it was first demonstrated that CD4 antibody selectively blocked syncytium formation (Dalgleish *et al.*, 1984). Finally, we use pseudotype viruses to determine functional entry and uncoating (McClure *et al.*, 1990; Takeuchi *et al.*, 1992; Vile *et al.*, 1991). Retrovirus 'pseudotypes' containing marker genes such as G418 resistance or β-galactosidase provide assays of infection that depend on the specificity of the retroviral envelope; likewise, vesicular stomatitis virus (VSV) pseudotypes bearing retrovirus glycoproteins provide an envelope-specific plaque assay. These pseudotypes have been of immense value in studying receptors, virus internalization and virus neutralization. Overall, a variety of assays of binding, fusion and functional entry may be used to determine different steps in the process of virus internalization.

On the surface of the retrovirus, the outer envelope glycoproteins (the surface or SU proteins) recognize the primary cellular receptor. The SU protein of HIV is called gp120 as it is a glycoprotein of approximately 120 kDa. Similarly, the SU protein, gp70, characterizes C-type murine retroviruses used as gene vectors. The transmembrane (TM) protein (gp41 for HIV) anchors the SU protein and mediates membrane fusion. SU–TM complexes are assembled on the envelope of retroviruses as trimers or tetramers according to the virus strain. They serve as the means by which the viral genes are delivered into the target cell.

2. The CD4 receptor for HIV

Working closely with Peter Beverley's immunology laboratory at University College London Medical School (UCLMS), we identified the CD4 antigen as the HIV receptor and mapped epitopes essential for viral recognition (Dalgleish *et al.*, 1984; Sattentau *et al.*, 1986). With the cloning of the cDNA for CD4, the binding site for HIV was mapped to the CDR2-like region on the N-terminal or first (D1) of four immunoglobulin-like domains of CD4 (Arthos *et al.*, 1989; Maddon *et al.*, 1986, 1988). For example, a single codon change at amino acid residue 43 entirely destroys the recognition and binding by gp120 of HIV without affecting the overall structure of the protein. As the D1 and D2 domains have been crystallized (Ryu *et al.*, 1990; Wang *et al.*, 1990), the effect of the mutations on the HIV binding site can be seen. There appear to be other regions of CD4 which are important after initial binding, with a temperature-dependent further fit involving conformational changes in both the cellular receptor molecule and the viral glycoproteins (Moore *et al.*, 1993).

The recognition of CD4 as the HIV receptor has led to a number of ideas for therapeutic intervention or prevention of HIV infection (Weiss, 1993). A soluble, recombinant form of the receptor molecule can be made by placing a stop codon before the transmembrane region. This protein has a very powerful neutralizing or inactivating effect on the virus. If it is fused with the Fc portion of an Ig molecule it attains a long plasma half-life in the human circulation. If the recombinant CD4 is linked or fused to a toxin then, upon binding to HIV-infected cells expressing gp120, the cells are killed. Ideas for HIV vaccines have also come from understanding gp120–CD4 interactions. For example, mirror-image anti-idiotypic antibodies to monoclonal antibodies, which recognize the same epitope on CD4 as does gp120, should neutralize the virus; hence the anti-CD4 monoclonal antibodies are potential vaccines. Furthermore, the regions of gp120 that bind to CD4 are conserved among all HIV-1 strains, and may therefore be the most suitable immunogens for a broad spectrum vaccine. However, anti-idiotypic antibodies and CD4 group-specific antibodies so far exert a very weak neutralizing effect compared to soluble CD4 or antibodies binding to the variable V3 loop of gp120 (*Table 1*). Thus there is a problem in combining potency in protection against infection with cross-neutralization of diverse HIV-1 and HIV-2 strains that would be required in a vaccine.

Knowing that CD4 is the receptor for HIV also helps to explain the pathogenesis of the virus leading to AIDS. The cells that become principally infected and depleted express CD4. These are T-helper lymphocytes, dendritic cells, macrophages and microglial cells in the brain. CD4 is expressed on T lymphocytes as part of the T-cell immune receptor complex. It interacts with class II major histocompatibility complex (MHCII) antigens in MHC-restricted reactions. The binding of gp120 to CD4 interferes with this interaction and is therefore immunosuppressive. If, however, the CD4 molecule is expressed without any of the other immune cell surface molecules in, say, HeLa cells, the CD4 antigen alone renders the cells susceptible to infection by HIV. In contrast, when the human CD4 gene is expressed in exactly the same construct in mouse cells, they remain resistant to infection by HIV. The CD4+ mouse cells bind the virus as there is plenty of human CD4 at the cell surface, but there is some block to cell fusion and entry as there is

Table 1. Specificity and potency of reagents that neutralize HIV

Virus		Neutralizing agent			
Strain	Origin	Anti-gp120 V3 loop	Anti-gp120 CD4 site	Anti-idiotype to anti-CD4	Recombinant soluble CD4
HIV-1					
LAI	France	400 000	512	8	10 000
SF-2	USA	0	1024	32	10 000
RF	Haiti	0	256	16	10 000
CBL4	Tanzania	0	256	16	10 000
JRCSF	USA	0	512	16	512
HIV-2 ROD	Guinea-Bissau	0	4	8	2 000
CBL22	Gambia	0	8	16	2 000

The values shown are the highest dilutions of 100 μg antibody or soluble CD4 that yield >90% neutralization of HIV infectivity. Thus anti-gp120 V3 loop (monoclonal antibody 110.5) will neutralize the LAI strain against which it was raised at 1:400 000 dilution, whereas it does not neutralize any other strains. The soluble CD4 receptor, however, neutralizes all strains, but the primary, non-fusigenic isolate JRCSF reacts more weakly than the syncytium-forming, laboratory strains (all the other strains shown).

no induction of syncytia or penetration of pseudotypes (Maddon *et al.*, 1986). Therefore, it appears that a second cell surface receptor is required which is on human cells but not murine cells. This receptor may not be particularly lineage specific as it is present on HeLa cells, B cells and T cells.

3. Retrovirus entry

How do retroviruses get into cells after binding to receptors at the cell surface? There are two main routes of virus entry. Some viruses initiate fusion at the cell surface so that the core of the virus can enter the cytoplasm directly and start its replication cycle. Paramyxoviruses, such as measles virus, enter cells in this way. On the other hand, influenza virus and many other viruses are taken up into clathrin-coated pits upon binding to their receptors. The coated pits form endocytotic vesicles or endosomes and this route of entry is called receptor-mediated endocytosis. Within the endosome, fusion between the viral envelope and the endosomal membrane occurs, to deliver the virus core to the cytoplasm.

With HIV one can see both routes of entry by electron microscopy. Cell surface fusion seems to be the regular route because infection can be obtained in the absence of receptor-mediated endocytosis, for example by mutating phosphorylation signals in the cytoplasmic domain of the CD4 receptor (Maddon *et al.*, 1988). Moreover, we studied receptor proteins that have the HIV-binding domain derived from human CD4 and the rest of the molecule derived from mouse CD4. Such chimeric receptors still work in human cells for HIV infection but still do not work in mouse cells, so a possible failure of human CD4 to interact with murine clathrin does not explain the resistance of murine cells to infection (Maddon *et al.*, 1988).

The pH in an endosome is about 5 rather than 7. The reason why influenza virus requires an endosomal means of entry is that the conformational changes in its outer envelope glycoprotein, haemagglutinin, are dependent on the low pH in the endosome to effect fusion. VSV is similarly pH dependent for fusion and entry. We therefore investigated pseudotypes of VSV bearing the envelope of HIV or other retroviruses. If the endosomal pH is increased by treating cells with drugs like monensin and chloroquine, or with NH_4Cl, VSV infection is inhibited. But if the VSV is wrapped up in the outer coat of HIV (the pseudotype), the plaque-forming efficiency of this virus in the presence of NH_4Cl compared to untreated cells is not impaired (*Figure* 2). Thus, VSV in its own envelope requires low pH to fuse with endosome membranes, whereas HIV-1, HIV-2 and human T-cell leukaemia virus are essentially pH independent for entry (McClure *et al.*, 1990). In fact the only retrovirus in *Figure 2* that seems to be pH sensitive is the ecotropic mouse leukaemia virus (MLV-E) and even that is not pH sensitive for XC cells. The reason behind this may provide a clue about HIV entry. It is not that low pH itself changes the conformation of the outer envelope glycoprotein for MLV-E, rather it is dependent on proteolytic cleavage of the surface glycoprotein which is mediated by an acid-dependent cathepsin in the endosome. The XC cells probably express a cell surface protease that cleaves gp70 at neutral pH. That is why MLV-E induces multinucleated syncytia in these cells too, whereas it does not in NIH-3T3 cells. Perhaps all the pH-independent retroviruses, including HIV, require proteolytic cleavage of the SU proteins too.

We made a series of cell lines besides murine cells that express human CD4. The transfected mink, rabbit, monkey and human cells

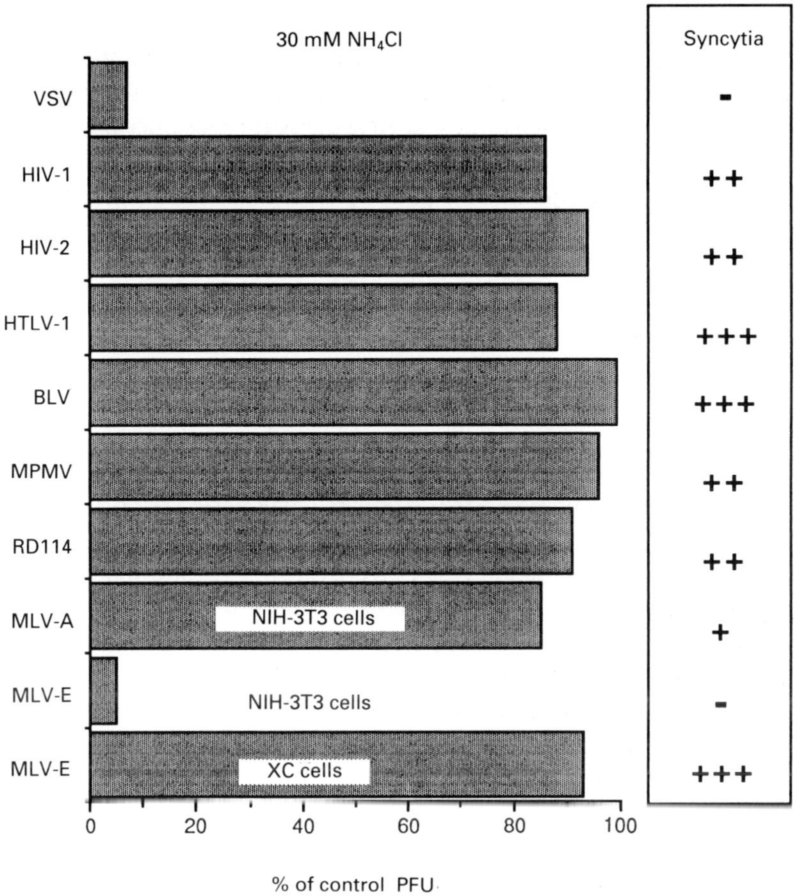

Figure 2. Sensitivity of VSV pseudotypes with retroviral envelopes to raised endosomal pH. Cells were treated with 30 mM NH_4Cl for 1 h before and 3 h after challenge with virus. The efficiency of plating was calculated from the VSV plaque titre compared to that on cells incubated in control medium. The cells were human sarcoma cells (HT1080) except where stated to be murine NIH-3T3 cells. The sensitivity to cell fusion on co-cultivation with cells producing the retrovirus is also indicated. Whereas VSV in its own envelope was dependent on low endosomal pH, only MLV-E among the retroviral envelope pseudotypes was pH dependent, for NIH-3T3 cells but not for XC cells. Data from McClure *et al.* (1990). Figure reproduced from Weiss *et al.* (1993) with permission from Plenum Publishing Corporation.

certainly express CD4 on the surface because binding of monoclonal antibodies to CD4 and recombinant gp120 from HIV can be detected. Yet most of these CD4-transfected cells are resistant to infection by

HIV-1 (Clapham *et al.*, 1991). They seem to be lacking a secondary factor or receptor expressed on human lymphocytes and HeLa cells.

The story becomes more complicated when different strains of HIV-1, HIV-2 and SIV, the simian immunodeficiency virus, are examined. Among the human cell lines expressing recombinant CD4 receptors, all are highly sensitive to HIV-2, but different cell lines are selectively sensitive to HIV-1 or SIV. Each of these three viruses binds to human CD4 on each cell type but it is not yet known what the second component is that seems to distinguish between HIV-1, HIV-2 and SIV. HIV-2 has a very broad host range and mink or cat cells can be infected provided that they express human CD4. Therefore transgenic animals might be useful as models for HIV-2 infection and pathogenesis (McKnight *et al.*, 1994).

One current hypothesis is that the second component is a cell surface protease (Clements *et al.*, 1991; Schulz *et al.*, 1993). As mentioned previously, cleavage of gp70 appears to be necessary for MLV-E infection. Furthermore, tryptic and chymotryptic cleavage sites have been identified in the V3 loop of HIV-1 gp120 and recombinant gp120 produced in CHO cells is often cleaved at the tryptic site. It is possible that, after adsorption to CD4, the entry of the virus particles is blocked unless the protease clips the V3 loop. This somehow leads to an overall conformational change that exposes the hydrophobic domain of the transmembrane protein gp41 to effect fusion between the virus envelope and the cell membrane. If a single codon mutation is made in gp120 of the HIV-1 LAI strain that prevents tryptic (thrombin) cleavage of the V3 loop, it renders virus rescued from the molecular clone non-infectious. However, other infectious stains of HIV-1 lack the thrombin cleavage site in V3. There may be different proteases that selectively cleave the envelopes of HIV-1, HIV-2 and SIV. A number of cell surface molecules have protease activity, including some of the CD differentiation antigens, for example CD13 and CD26. Neutralizing antibodies that bind to the V3 loop do block cleavage by thrombin, so these very potent but rather type-specific neutralizing antibodies (*Table 1*), which neutralize the virus after it has bound to cell surface CD4, may block infection by blocking proteolytic cleavage. One recent report which found that CD26 acts as the second receptor (Callebaut *et al.*, 1993) has not been confirmed in other laboratories, including ours (Patience *et al.*, 1994).

4. HIV receptors and pathogenesis

Do these studies on the infection of CD4-transfected cell lines *in vitro* have any relevance to the pattern of infection by HIV in infected people? The cell types that appear to be infected *in vivo* are principally peripheral blood lymphocytes, that is, the CD4+ subset of T-helper lymphocytes, and also CD4+ cells of the monocytic lineage, including tissue macrophages and the microglia of the brain. Peripheral blood dendritic cells and the related Langerhans cells in the skin are also sensitive to HIV infection via CD4 receptors. However, different strains of HIV have subtly different tropisms. Whereas most of them grow in peripheral blood lymphocytes, there are some that will specifically grow in macrophages that cannot grow in established T-cell leukaemia lines, and some that grow in T-cell lines that cannot infect macrophages. In some cases the V3 loop epitope in gp120 can be transferred from a macrophage tropic strain into a T-cell line tropic strain of HIV to obtain a switch in tropism (Hwang *et al.*, 1991). We selected a mutant of a cloned primary HIV isolate to expand its host range from peripheral blood lymphocytes only to growing in certain T-cell lines. The crucial mutation, however, is in a different part of the gp120, in the V1 loop (Boyd *et al.*, 1993). In both these cases, as in our studies of CD4-transfected cell lines, the cellular tropism of the virus appears to be determined at a post-adsorption, pre-penetration stage of infection.

Although CD4 appears to be essential for HIV infection of lymphocytes, alternative receptors may play a role in other cell types (Weiss, 1993). There is evidence for binding of HIV gp120 to glycolipids with terminal galactosyl-ceramide residues (Bhat *et al.*, 1991) and for infection via Fc receptors of HIV opsonized by antibodies (McKeating *et al.*, 1990). HIV-2 may enter certain CD4– human cell types provided its envelope is first 'activated' by soluble CD4 (Clapham *et al.*, 1992).

5. Receptors for other retroviruses

We have used similar techniques to those used for HIV and CD4 to study human receptors for other retroviruses. We classified how many different kinds of retrovirus receptors there might be on human cells by mixing pairs of cell lines chronically infected with different virus strains (Sommerfelt and Weiss, 1990). With all their receptors saturated, no syncytia are formed, but when viruses using different receptors are mixed together they cross-fuse each other and massive syncytia are seen. Taking a panel of 25 different strains of human and animal retroviruses, it was possible to classify them into eight different receptor groups (*Table 2*). These results were confirmed using VSV pseudotypes and retroviral vectors carrying marker genes such as histidinol metabolism or β-galactosidase (Sommerfelt and Weiss, 1990; Takeuchi *et al.*, 1992).

For gene therapy targeted to stem cells and lymphocytes, the group 1 receptor or the group 5 receptor might prove to be as good or a better receptor for entry than the amphotropic receptor (group 3). One surprise was that two quite unrelated leukaemia retroviruses from gibbon apes (GALV) and cats (FeLV-B), with no obvious homology in their surface glycoproteins, use the same receptor (Takeuchi *et al.*, 1992).

Figure 3 shows that the molecular natures of the recently discovered

Table 2. Receptor groups of 25 retroviruses infecting human cells

Receptor group	Virus	Type	Host
1	SRV-1	Type D	Rhesus macaque
	SRV-2	Type D	Black macaque
	MPMV	Type D	Rhesus macaque
	SRV-4	Type D	Cynomolgus macaque
	SRV-5	Type D	Rhesus macaque
	PO-1-Lu	Type D	Spectacled langur
	SMRV	Type D	Squirrel monkey
	BaEV	Type C	Baboon
	RD114	Type C	Domestic cat
2	MLV-A	Type C	Mouse
3	MLV-X	Type C	Mouse
4	FeLV-C	Type C	Domestic cat
5	FeLV-B	Type C	Domestic cat
	SSAV	Type C	Woolly monkey
	GALV	Type C	Gibbon ape
6	BLV	Tax-Rex	Cattle
7	HTLV-1	Tax-Rex	Human
	HTLV-2	Tax-Rex	Human
	ChTLV	Tax-Rex	Chimpanzee
	STLV	Tax-Rex	Bonnet macaque
8	HIV-1	Lenti	Human
	HIV-2	Lenti	Human
	SIV_{mac}	Lenti	Rhesus macaque
	SIV_{smm}	Lenti	Sooty mangabey
	SIV_{agm}	Lenti	Green monkey

Data from Sommerfelt and Weiss (1990).

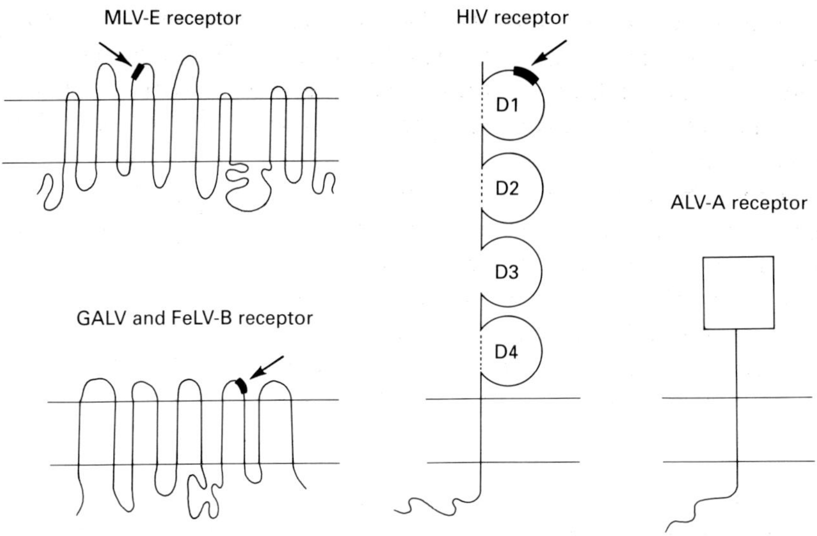

Figure 3. Schematic configuration of the known retrovirus receptor molecules. The MLV-E receptor spans the lipid bilayer 14 times and functions as a basic amino acid transporter (Wang *et al.*, 1991; see Vile and Weiss, 1991). The GALV and FeLV-B receptor (Takeuchi *et al.*, 1992) resembles a phosphate transporter (Johann *et al.*, 1992). The CD4 receptor is a member of the immunoglobulin superfamily (Maddon *et al.*, 1986, 1988). The receptor for bovine leukosis virus (BLV) is a previously unknown protein (Ban *et al.*, 1993). The receptor for avian leukosis and sarcoma viruses, subgroup A (ALV-A) is related to the low density lipoprotein receptor (LDLR) family (Bates *et al.*, 1993). The amphotropic receptor is related to the GALV/FeLV-B receptor (Miller *et al.*, 1994). Arrows show the determining domains for viral recognition. Adapted from Weiss (1993) with permission from Plenum Publishing Corporation.

retrovirus receptors appear to be completely different. What might their natural functions be? The CD4 receptor interacts with MHCII antigens in MHC-restricted immune reactions; indeed the binding of high doses of HIV gp120 can block T-cell activation. The MLV-E receptor is an amino acid transporter for basic amino acids (Kim *et al.*, 1991; Wang *et al.*, 1991), and the GALV–FeLV-B receptor may be a phosphate transporter because the only protein even distantly homologous in the data banks is a phosphate transporter of the mould *Neurospora crassa* (Johann *et al.*, 1992).

We have mapped the determining site for both GALV and FeLV-B to one small region of the receptor. The epitope necessary for virus recognition is surprisingly varied in mammalian species, even between rats and compared to the rest of the protein sequence. The rat receptor is susceptible to GALV and not to FeLV-B, whereas the human receptor reacts to both and the mouse receptor to neither. We found that single codon mutations within the critical epitope of the receptor change its recognition of GALV or FeLV-B (Tailor *et al.*, 1993). We do not yet know whether this is a primary receptor for virus attachment or a secondary receptor mediating membrane fusion.

Other human receptor genes have been mapped to specific chromosomes but we do not yet know what the receptors are (Sommerfelt *et al.*, 1988, 1990). The amphotropic MLV receptor has recently been identified as one related to the GALV receptor (Miller *et al.*, 1994). There are many reasons why retrovirus receptors will continue to be of interest: for elucidating mechanisms of entry, for determining cell and tissue tropisms, for understanding viral pathogenesis and for gene therapy.

Acknowledgements

I would like to acknowledge my colleagues Paul Clapham, Thomas Schulz and Myra McClure, who have been involved in the HIV studies and pH-sensitive entry; our work with other receptors and retroviral vectors was largely done by three students, Maja Sommerfelt, Richard Vile and Chetankumar Tailor, and by Yasuhiro Takeuchi as a joint project with Mary Collins' laboratory. Our research is supported by the Medical Research Council and the Cancer Research Campaign.

References

Arthos J, Deen CK, Chaikin MA *et al.* (1989) Identification of the residues in human CD4 critical for the binding of HIV. *Cell,* **57**, 469–481.

Ban J, Portelle D, Altaner C, Horion B, Krchnak V, Burny A and Kettmann R. (1993) Isolation and characterisation of a 2.3-kilobase-pair cDNA fragment encoding the binding domain of the bovine leukosis virus cell receptor. *J. Virol.* **67**, 1050–1057.

Bates P, Young JAT and Varmus HE. (1993) A putative receptor for subgroup A Rous sarcoma virus is related to the LDL receptor. *Cell,* **74**, 1043–1051.

Bhat S, Spitalnik SL, Gonzalez-Scarano F and Silberberg DH. (1991) Galactosyl ceramide or a molecule derived from it is an essential component of the neural receptor for HIV-1 envelope glycoprotein gp120. *Proc. Natl Acad. Sci. USA,* **88**, 7131–7134.

Boyd MT, Simpson GR, Cann AJ, Johnson MA and Weiss RA. (1993) A single amino acid substitution in the V1-loop of human immunodeficiency virus type 1 gp120 alters cellular tropism. *J. Virol.* **67**, 3649–3652.

Callebaut C, Krust B, Jacotot E and Hovanessian AG. (1993) T cell activation antigen, CD26, as a cofactor for entry of HIV in $CD4^+$ cells. *Science*, **262**, 2045–2050.

Clapham PR, Blanc D and Weiss RA. (1991) Specific cell surface requirements for infection of CD4-positive cells by human immunodeficiency virus type 1, type 2 and simian immunodeficiency virus. *Virology*, **181**, 703–715.

Clapham PR, McKnight A and Weiss RA. (1992) Human immunodeficiency virus type 2 infection and fusion of CD4-negative human cell lines: induction and enhancement by soluble CD4. *J. Virol.* **66**, 3531–3537.

Clements GJ, Price-Jones M, Stephens PE *et al.* (1991) The V3 loops of the HIV-1 and HIV-2 surface glycoproteins contain proteolytic cleavage sites: a possible function in viral fusion. *AIDS Res. Hum. Retrovir.* **7**, 3–16.

Dalgleish AG, Beverley PCL, Clapham PR, Crawford DH, Greaves MF and Weiss RA. (1984) The CD4 (T4) antigen is an essential component of the receptor for the AIDS retrovirus. *Nature*, **312**, 763–766.

Hwang SR, Boyle TJ, Lyerly K and Cullen BR. (1991) Identification of the envelope V3 loop as the primary determinant of cell tropism in HIV1. *Science*, **253**, 71–74.

Johann SV, Gibbons JJ and O'Hara B. (1992) GLVR1, a receptor for gibbon ape leukemia virus, is homologous to a phosphate permease of *Neurospora crassa* and is expressed at high levels in the brain and thymus. *J. Virol.* **66**, 1635–1640.

Kim JW, Closs EI, Albritton LM and Cunningham JM. (1991) Transport of cationic amino acids by the mouse ecotropic retrovirus receptor. *Nature*, **352**, 725–728.

Maddon PJ, Dalgleish AG, McDougal JS, Clapham PR, Weiss RA and Axel R. (1986) The T4 gene encodes the AIDS virus receptor and is expressed in the immune system and brain. *Cell*, **47**, 333-348.

Maddon PJ, McDougal JS, Clapham PR, Dalgleish AG, Jamal S, Weiss RA and Axel R. (1988) HIV infection does not require endocytosis of its receptor, CD4. *Cell*, **54**, 865–874.

McClure MO, Sommerfelt MA, Marsh M and Weiss RA. (1990) The pH independence of mammalian retrovirus infection. *J. Gen. Virol.* **71**, 767–773.

McKeating JA, Griffiths PD and Weiss RA. (1990) HIV susceptibility conferred to human fibroblasts by cytomegalovirus-induced Fc receptor. *Nature*, **343**, 659–661.

McKnight A, Clapham PR and Weiss RA. (1994) HIV-2 and SIV infection of non-primate cell lines expressing human CD4: restrictions to replication at distinct stages. *Virology*, in press.

Miller DG, Edwards RH and Miller AD. (1994) Cloning of the cellular receptor for amphotropic murine retroviruses reveals homology to that for gibbon ape leukemia virus. *Proc. Natl Acad. Sci. USA*, **91**, 78–82.

Moore JP, Jameson BA, Weiss RA and Sattentau QJ. (1993) The HIV–cell fusion reaction. In: *Viral Fusion Mechanisms* (ed. J Bentz). CRC Press,

Boca Raton, FL, pp. 233–289.

Ryu SE, Kwong PD, Truneh A *et al.* (1990) Crystal structure of an HIV-binding recombinant fragment of CD4. *Nature*, **348**, 419–426.

Sattentau QJ, Dalgleish AG, Weiss RA and Beverley PCL. (1986) Epitopes of the CD4 antigen and HIV infection. *Science*, **234**, 1120–1123.

Schulz TF, Reeves J, Hoad JG, Tailor C, Stephens P, Page K, Moore JP and Weiss RA. (1993) Effect of mutations in the V3 loop of HIV-1 gp120 on infectivity and susceptibility to proteolytic cleavage. *AIDS Res. Hum. Retrovir.* **9**, 159–166.

Sommerfelt MA and Weiss RA. (1990) Receptor interference groups among 20 retroviruses plating on human cells. *Virology*, **176**, 58–69.

Sommerfelt MA, Williams BP, Clapham PR, Solomon E, Goodfellow PN and Weiss RA. (1988) Human T cell leukemia viruses use a receptor determined by human chromosome 17. *Science*, **242**, 1557–1559.

Sommerfelt MA, Williams BP, McKnight A, Goodfellow PN and Weiss RA. (1990) Localization of the receptor gene for type D simian retroviruses on human chromosome 19. *J. Virol.* **64**, 6214–6220.

Tailor C, Takeuchi Y, O'Hara B, Weiss RA and Collins MKL. (1993) Mutation of amino acids within the gibbon ape leukaemia virus (GALV) receptor differentially affects feline leukaemia virus subgroup B (FeLV-B), simian sarcoma-associated virus (SSAV) and GALV infection. *J. Virol.* **67**, 6737–6741.

Takeuchi Y, Vile RG, Simpson G, O'Hara B, Collins MKL and Weiss RA. (1992) Feline leukemia virus subgroup B uses the same cell surface receptor as gibbon ape leukemia virus. *J. Virol.* **66**, 1219–1222.

Vile R and Weiss RA. (1991) Virus receptors as permeases, *Nature*, **352**, 666–667.

Vile R, Schulz TF, Danos O, Collins MKL and Weiss RA. (1991) A murine cell line producing HTLV-1 pseudotype virions carrying a selectable marker gene. *Virology*, **179**, 420–424.

Wang J, Yan Y, Garrett TPJ *et al.* (1990) Atomic structure of a fragment of human CD4 containing two immunoglobulin-like domains. *Nature*, **348**, 411–419.

Wang H, Kavanaugh MP, North RA and Kabat D. (1991) Cell surface receptor for ecotropic murine virus is a basic amino-acid transporter. *Nature*, **352**, 729–731.

Weiss RA. (1993) Cellular receptors and viral glycoproteins involved in retrovirus entry. In: *The Retroviridae*, vol. 2 (ed. JA Levy). Plenum Press, New York, pp. 1–108.

10

Viral vectors for gene therapy

G.W.G. Wilkinson[1], R.L. Darley[2] and P.R. Lowenstein[3]
Departments of [1] Medicine and [2] Haematology, University of Wales College of Medicine, Cardiff, UK
[3] Department of Physiology, University of Wales College of Cardiff, Cardiff, UK

1. Introduction

The ability of viral vectors to act as vehicles for efficient gene transfer has provided the basis on which gene therapy has been developed. Gene therapy can be defined as the direct transfer of recombinant DNA into a patient's cells to treat a disorder. Viruses, in contrast, directly transfer their genetic material into cells and, as a consequence, can be the cause of a disorder. The conversion of erstwhile pathogens into vectors for therapeutic benefit has also provided virologists with unexpected problems and fresh challenges. Gene therapy can, however, learn much from nearly 200 years experience of using defined live viral vaccines to control and eradicate some of the major diseases of mankind.

The demands made by a gene therapy on its vector are extreme. Vectors have to be able to access the target cell population, introduce the DNA into the cell in a transcriptionally active state, be non-toxic and be able to express the therapeutic gene product at sufficient levels and for an appropriate length of time to correct the disease phenotype. To overcome the toxic consequences of virus infections, vectors are usually designed to be not only replication-defective but also to express only the transgene and not viral proteins. While viral vectors may be considered simply as highly efficient gene delivery systems, some also provide a mechanism for sustained expression in target cells.

The excitement of gene therapy lies in its potential for widespread clinical application. Although the most obvious use is in the treatment of a genetic metabolic defect (such as haemophilia or cystic fibrosis), perhaps the greatest potential of the technology lies in the therapeutic treatment of disease without the need for a defined underlying genetic cause. A number of chapters in this volume cover gene therapy approaches to cancer treatment, and strategies are being considered for a diverse range of ailments, such as cardiovascular disease, wound repair and arthritis. Indeed, virus infection can be considered to constitute a genetic disease and gene therapy protocols have been developed for the management of human immunodeficiency virus (HIV). An extremely liberal interpretation could consider more conventional vaccination (e.g. with a Vaccinia recombinant) to be a form of gene therapy. Furthermore, a significant number of approved protocols are for gene marker, rather than gene therapy, experiments, where the function of the transgene is to follow the fate of specific cell populations, often in bone marrow transplants.

Clearly, gene delivery and expression systems have to be selected, developed and tailored to meet the needs of a vast range of potential treatment protocols. Progress is currently severely constrained by the limitations of the available technology to effectively deliver, control and maintain expression of therapeutic transgenes. Although there are nearly as many viral expression systems as there are characterized viruses, this chapter will concentrate on the mainstream candidate vectors currently being developed specifically for gene therapy. Attention is also drawn to the value of characterized viral genetic elements in hybrid expression systems and, in particular, as potential components of direct DNA transfer systems (Section 6).

2. Retroviruses

Retroviruses historically have been the subject of an intense research effort because of their associated oncogenicity, both at high frequency, with viruses encoding oncogenes, and at low frequency, by insertional mutagenesis. Additionally, the emergence of AIDS has generated a wealth of information on the biology of the *Lentivirus* sub-family. The majority of vectors used in gene therapy protocols are, however, based on murine retroviruses. Retroviral vectors have been developed specifically for gene therapy and, consequently, represent the best characterized system. This has been an important factor in reassuring the legislative authorities over safety and, now that this technology has been applied in the clinic over a number of years, the use of retroviruses for gene therapy is becoming more acceptable (Tolstoshev, 1993). The reasons why this group of vectors are particularly suited to gene therapy are rooted in the molecular biology of the retrovirus replicative cycle.

2.1. Retroviral life cycle

A typical retrovirus particle has a ribonucleoprotein core, consisting of two copies of genomic RNA, the structural core proteins (encoded by the *gag* gene), a reverse transcriptase (encoded by *pol*) and an envelope containing the glycoprotein (encoded by *env*) which can determine the host range of the virus. Upon infection, the virus binds its receptor on the cell surface and is internalized (*Figure 1*). The RNA genome is reverse-transcribed to produce a double-stranded DNA molecule which migrates to the nucleus. The double-stranded DNA molecule is then integrated into host DNA by virtue of an integrase function associated with the *pol* gene product. A single viral particle generates a single stably integrated provirus encoding the *gag*, *pol* and *env* genes flanked by long-terminal repeat (LTR) sequences which contain the *cis*-acting elements necessary for transcribing and replicating the viral genome. The viral proteins, gag and pol, are translated from a full-length genomic transcript; the env protein is translated from a sub-genomic RNA which arises through splicing out of the *gag* and *pol* sequences. The full-length genomic RNA molecule also contains a recognition sequence, ψ, which constitutes a virion packaging signal. To complete the life-cycle, the retroviral genome is transcribed, packaged and the newly produced retroviral particles released non-destructively by budding from the cell surface.

2.2. Retroviral vectors for gene therapy

A remarkable feature of the retrovirus life-cycle is that infection, reverse transcription, integration and transcription all occur in the absence of any *de novo* viral gene expression (*Figure 1*). Consequently, all these processes can occur even if all coding elements of the virus are deleted. This makes it possible to generate vectors in which all viral protein coding sequences can be deleted and replaced with the gene(s) to be expressed. These artificial 'genomes' can be packaged to form infectious viral particles by transfecting the vector DNA into a cell line which expresses the viral structural proteins, i.e. gag, pol and env. These are called packaging cells (or sometimes helper cells), because they comple-

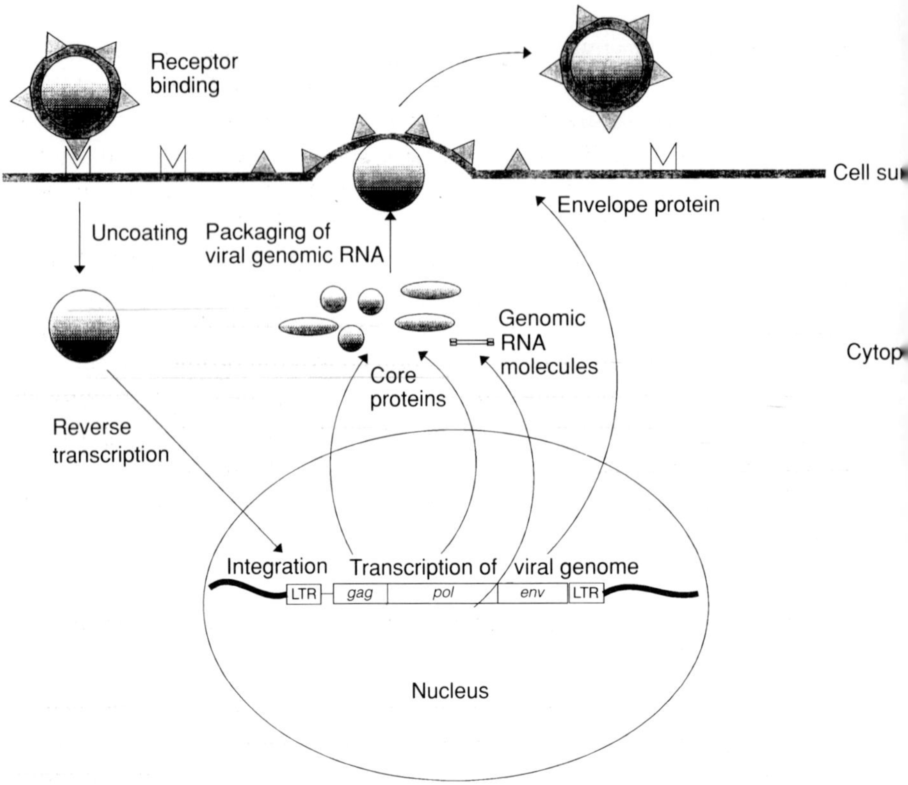

Figure 1. Retroviral life-cycle (see text for details).

ment, in *trans*, the deficiencies of the engineered vector. The host range of the packaged virus is determined by the type of env protein expressed by the packaging line; amphotropic packaging lines generate virus able to infect a wide range of human tissues, while virus packaged by ecotropic packaging cells have a narrower host range, which does not include human cells. It is most important that the retroviral genome providing the helper functions is not itself packaged. To ensure this does not happen, the packaging signal (ψ) in the helper genome is deleted (see also Section 2.3). The encapsulated recombinant is infectious, but is not capable of self-replication and, therefore, provides 'one-hit' gene delivery. The recombinant genome is reverse-transcribed and stably integrated, as for the wild-type virus. Chromosomal integration is a crucial feature of this vector as it allows the transgene to be inherited by subsequent generations derived from the transduced cell. Since in the commonly used vector systems the integrated DNA cannot express any viral proteins, there are no associated detrimental effects on the transduced cell, for example through *trans*-activation of host genes or by stimulation of an immune response against the infected cell.

Retroviral vectors are highly adaptable and a great variety have been developed which utilize different strategies for gene expression (see Miller, 1992, for review). Vectors employed in gene therapy are mostly based on Moloney murine leukaemia virus (MoMLV) and are engineered to express either a single gene, or two different genes (e.g. a therapeutic gene and a marker gene) by utilizing both the *gag/pol* and *env* positions of the retroviral genome. Expression of the first or single gene is usually driven by the endogenous retroviral promoter (the LTR). Expression of a second gene can also be driven by the LTR by relying on the natural splicing mechanism of the retrovirus, or the second gene can be cloned in with its own promoter (commonly the SV40 or herpes simplex *tk* gene promoters) to drive expression without the need for splicing.

2.3. Limitations and potential problems with using retroviruses for gene therapy

The biology of retroviruses imposes certain limitations on their use in gene therapy. Integration of murine retroviral vectors into chromosomal DNA is dependent on mitosis and, since no expression occurs without integration, non-dividing tissues (e.g. neuronal tissue) cannot be targeted, although in one therapeutic protocol this property has been turned to advantage (see Section 2.5). The retroviral genome does not have to be of a particular size to be packaged but there is an upper limit of 10–11 kb, restricting insert size to around 9 kb. Near the upper limit, there may also be a reduction in titre as the genome is packaged less efficiently. Retroviral vectors, consequently, cannot accomodate exceptionally large genes or particularly long promoter/regulatory elements.

Viral vectors have to be able to infect the target cells efficiently. The infectivity of the target can vary, being dependent on how actively the cell is cycling or on the density of retroviral receptors. However, the titre of retrovirus produced by packaging lines is rarely higher than 10^6 infectious units/ml and this is often a limiting factor. The instability of the retroviral particles following budding from the packaging cell also limits the potential titre. In an interesting experimental system, substituting the retrovirus envelope with that of vesicular stomatitis virus has permitted concentration of the 'pseudotype' virus by centrifugation to enable significantly higher titres to be achieved (Burns *et al.*, 1993). Nevertheless, a major limitation of retroviral gene therapy at present is the ability to infect sufficient numbers of target cells.

The random way in which the retrovirus provirus integrates into the host genome has two undesirable consequences. The first is that the level of retroviral gene expression may be highly variable. This is illustrated in *Figure 2*, in which the expression of the *lacZ* reporter gene is assayed in a population of infected fibroblasts. Although it may be possible to switch transgene expression on or off using conditional promoters, quantitative control may not be realistic. For gene therapy, retroviral vectors tend to be applied where the expression level is not critical to obtaining therapeutic benefit (e.g. 10–20% production of the normal levels of factor IX can alleviate severe haemophilia B). Overexpression of this factor, on the other hand, would not be expected to cause problems. The second problem is that of insertional mutagenesis. This term is used to describe the possible adverse consequences of insertion of the retroviral vector upstream of a proto-oncogene (causing its overexpression) or the disruption of expression of a tumour suppressor gene. A statistical analysis has been carried out on the possibility of cancer arising through such an event (Moolten and Cupples, 1992) and the risks were found to be very small. It should be noted that, experimental conditions in which immunosuppressed animals were chronically infected by replication-competent virus have resulted in malignancy; however, it is calculated that these animals were exposed

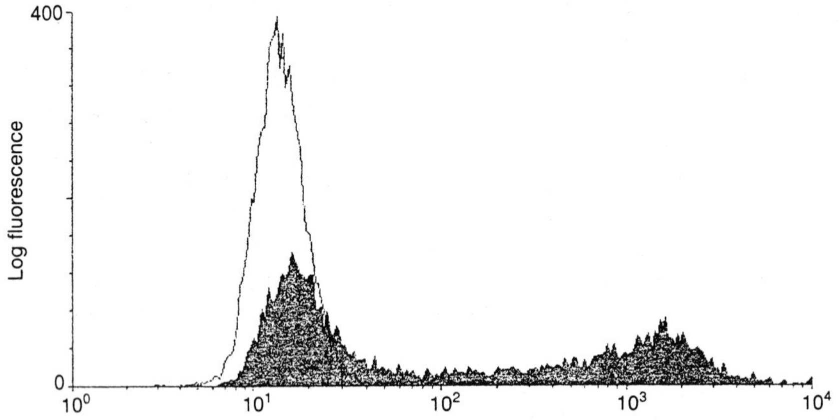

Figure 2. Fluorescence histograms illustrating the range of β-galactosidase expression in fibroblasts infected with a retrovirus (M7J-Z) bearing the *lacZ* gene (filled histogram). Background expression in mock-infected cells is shown by the open histogram. Cells were assayed by flow cytometry following exposure to the fluorogenic substrate, fluorescein-di-β-galactopyranoside.

to very high titres of virus and such a situation is very unlikely to arise in gene therapy protocols (Anderson, 1993).

This safety issue raises the question of the possibility of recombination event(s) leading to the generation of a replication-competent helper virus. The principal risk here is thought to lie with the packaging lines (for review, see Miller, 1990). Certainly, the first generation packaging lines (e.g. ψ-2 and ψ-AM), which had a simple deletion of the packaging signal, required only one recombination event with the vector construct to regenerate replication-competent virus. The current generation of packaging lines (e.g. GP+E86 and GP+*env*AM12) express the gag and pol proteins from one inserted provirus and the env proteins from a second construct. Additionally, deletions and mutations have been introduced into these helper constructs so that three separate recombination events would be required to generate replication-competent virus. So far, there have been no reports of replicating virus arising from these packaging lines. The possibility that vector-producing cells could package endogenous sequences (e.g. VL30), or that recombination could occur with other integrated proviruses, are concerns that are more difficult to allay. However, a note of comfort is that human complement inactivates murine retroviruses in a non-antibody-dependent manner, including those with an amphotropic envelope protein (Cornetta *et al.*, 1990). A comprehensive consideration of the safety issues and the problems with detecting any adverse consequences of gene therapy is covered by Temin (1990) and Ledley *et al.* (1992).

Gene expression from MoMLV-based retroviral vectors can be achieved in a broad range of cell types, although this is not always a desirable property since it makes it difficult to target retroviral expression to a particular tissue, unless the infection can be carried out *ex vivo*. A way of overcoming this difficulty is to base the vector on a retrovirus which has a more restricted tropism, for instance vectors based on Friend MLV show high-level expression in erythroid cells, but not in lymphoid cells, whereas MoMLV-based vectors express in both lineages (Bösze *et al.*, 1986). Tailoring the type of vector used to the intended target tissue can also improve levels of expression; for example, vectors based on the myeloproliferative sarcoma virus have been shown to be more active in myeloid cells than vectors based on MoMLV (Bowtell *et al.*, 1988). There is also evidence that expression from MoMLV-based vectors may be silenced in the ultimately desirable target for gene therapy, the stem cell; this is apparently due to the presence of sequences within the LTR which act as recognition elements for transcriptional repressor factors (Kempler *et al.*, 1993). Additionally, there appears to be a fundamental block to proviral integration in some stem cells, which affects a variety of different types of vector (Beck-Engeser *et al.*, 1991).

2.4. Target tissues for retroviral gene therapy

Most applications of retroviruses to gene therapy target haematopoietic cells. Perhaps the most significant reason for this is that infection can be carried out *ex vivo*. This overcomes the problem of the low infectivity of retroviruses, since the infection can be carried out in an intensive way, i.e. the cells can be co-cultivated with large volumes of retroviral supernatant and could undergo multiple rounds of infection in the presence of polycations, which enhance infection. Additionally, haematopoietic cells can be made to cycle actively *in vitro* by stimulation with growth factors. Following infection, the efficacy of the treatment can be assessed and checks can be made to ensure the growth potential of the transduced cells before they are reintroduced into the patient. After the introduction of the treated cells into the patient, the accessibility of this tissue facilitates subsequent monitoring of their fate *in vivo*.

Gene therapy of haematological disease has become even more attractive now that the pluripotent stem-cell fraction can be relatively easily identified by monoclonal antibodies. These stem cells occur in such low frequency in normal bone marrow that transducing significant numbers would be difficult in unfractionated populations. However, the ability to infect purified populations raises the realistic possibility that, if stable expression can be maintained, a single course of gene therapy would be sufficient for the lifetime of the individual. This type of treatment would also represent the lowest risk by minimizing the number of insertional events in the patient's cells. Transgene expression would be expected to occur in all cell lineages derived from the infected stem cell, which could, in some circumstances, be undesirable. Stem cells, for the most part, are expected to be quiescent in normal marrow, so there is some concern that stimulation with growth factors to induce cycling may enhance the efficiency of gene transfer at the expense of maintaining the self-renewing cell phenotype. Currently, there is much interest in the repopulating stem cells found in the peripheral blood of patients undergoing regenerative haematopoesis following cytotoxic drug therapy. Stem cells isolated from these patients have been found to be optimal targets for gene therapy, presumably because a high proportion are proliferating. For the same reason, stem cells derived from cord blood cells have also been found to be good targets for gene transfer using retroviruses. Another major target has been tumour cells (e.g. in melanomas and brain tumours), which are ideal for retroviral gene therapy since they are likely to be actively proliferating

2.5. Clinical applications of retroviral vectors

Gene marking with retroviruses. This is not strictly gene therapy, however a large proportion of the current clinical protocols are based on this type of application. In one type of gene tagging protocol, bone marrow harvested from leukaemia patients in remission has been transduced with the neomycin resistance gene (Brenner *et al.*, 1993). When the transduced marrow is used to reconstitute the haematopoietic system after ablative chemotherapy, a subsequent relapse in which the malignant cells were 'marked' would indicate the source of the residual disease to be from the autograft. Unmarked malignant cells would indicate that residual disease in the patient survived the cytotoxic therapy. Since the levels of residual disease in remission marrow may be very low, this does rely on a significant proportion of the marrow being marked. It is hoped that such studies will enable more effective strategies to be developed for treatment of myeloblastic leukaemia. Gene marking of tumour-infiltrating lymphocytes (TILs) has also been carried out. TILs have shown some promise in the treatment of melanoma and renal cell cancer; however, only a proportion of the patients respond (35–40%). In order to monitor the efficacy of TIL therapy, therefore, these cells have been marked so that their ability to home to the tumour tissue could be monitored (Aebersold *et al.*, 1990).

Gene therapy with retroviruses. Retroviruses were the first vectors to be used for gene therapy and have so far been applied in two principal areas: first, for the treatment of rare inherited genetic disorders where the lesion has been characterized and, second, for the treatment of cancer (for review, see Morgan and Anderson, 1993). The treatment of genetic disorders is based on the introduction of a functional cellular gene to complement a missing or defective gene, for example in the treatment of severe combined immune deficiency (SCID), the gene encoding adenine deaminase is transferred. This has met with some success for the first patient to be treated in this way in 1990; after eight infusions of modified cells, T cell numbers have increased 10-fold, tonsil regrowth has occurred and there has been a decreased number of infectious complaints. Other patients are showing a similarly promising outcome (Blaese, 1993). See Chapter 3 for further details of the use of gene therapy to treat SCID and other X-linked immunodeficiency diseases.

A number of approaches have been made in the treatment of cancer, both through direct modification of the tumour to induce its destruction (or suppress its proliferation) and by indirectly promoting tumour destruction by stimulation of the anti-tumour immune response. The latter can be achieved either by modifying the tumour cells themselves to enhance their immunogenicity or by modification of immune effector cells (e.g. tumour-infiltrating lymphocytes). In both cases, the target cells are infected *ex vivo* with cytokine genes (e.g. IL-2 or -4) to augment immune activity.

Infection of tumour cells *in situ* using retroviruses is technically difficult, firstly because of the problem of infecting a sufficient proportion of the tumour; here the likely limiting factors are retroviral titre and distribution of the retrovirus throughout the tumour mass. The second problem is that normal cells may also be modified when retroviruses are introduced *in vivo*. Despite these difficulties, this type of therapy has been applied in some situations; one particularly promising usage is in the treatment of cerebral gliomas. In this case, the problem of retroviral titre is overcome by injecting the packaging cells themselves, which secrete a retrovirus recombinant encoding the thymidine kinase gene. Normal brain tissue, being composed of non-dividing cells, will not be transduced, but both the tumour cells and the injected packaging cells will express thymidine kinase, which renders them susceptible to the cytotoxic effects of nucleoside analogues (such as Ganciclovir). Thus, after allowing a period for infection to take place, both the tumour tissue and the packaging cells can be killed by administration of the drug. The effectiveness of this treatment is apparently augmented by a 'bystander effect', in which tumour cells which have not been modified are also killed by the treatment – the mechanism of this phenomenon is not understood, but suggests that, at least in some cases, it may not be necessary to modify all the cells in a tumour to eliminate disease.

Retroviruses have also been employed to deliver anti-sense sequences to tumour cells specific for activated oncogenes (e.g. *ras*) in an attempt to arrest proliferation. Alternatively, the same result may be achieved by introduction of tumour suppressor genes (e.g. *p53*). A full discussion of gene therapy of cancer can be found in Chapter 8. Considerable interest has been generated in a range of ingenious gene therapy strategies aimed at preventing HIV infection and these are the subject of an excellent recent review (Yu *et al.*, 1994).

3. Adenovirus vector systems

Recent developments with adenovirus (Ad) recombinants have done much to engender the belief that *in vivo* gene therapy is a credible enterprise. Ad, itself, is normally associated with mild or asymptomatic infection of the respiratory tract, keratoconjunctivitis and gastroenteritis but can, albeit rarely, be the cause of broncho-pneumonia in infants and systemic infection in immunocompromised individuals. The common serotypes are endemic childhood infections in developed countries (e.g. serotypes 1, 2, 5) while the less prevalent serotypes can be associated with sporadic outbreaks of disease (Horwitz, 1990). In order to prevent episodes of acute respiratory disease (ARD) debilitating new recruits, a live enteric-coated vaccine was developed for the US military, based on Ad types 4 and 7, and has been used successfully over a 20-year period. This programme provides an encouraging background for the development of recombinant Ad for both vaccination and gene therapy. A significant effort has been put into vector development. Replication-competent Ad vectors have been used to provide high level expression of cloned gene products for many years and have been an effective vehicle for vaccine delivery in experimental systems (Berkner, 1988; Graham and Prevec, 1991; Grunhaus and Horwitz, 1992; Wilkinson, 1994). However, it is the emergence of replication-defective vectors as a high efficiency gene delivery system that has provided an option for *in vivo* gene therapy. Although the utility of replication-defective Ad vectors has exceeded all expectations, their limitations are also now beginning to be appreciated.

3.1. Replication-defective Ad vectors

The Ad genome is a linear doubled-stranded DNA molecule of approximately 36 kb which circularizes following infection and is transcribed, replicated and packaged into viral particles (*Figure 3*) within the nuclei of infected cells. The Ad E1 gene region is essential, not only for replication, but also for activating all virus-encoded early and then late phase gene expression (*Figure 4*). Replication-defective vectors based on E1 deletion mutants can, however, be propagated on a cell line expressing an integrated copy of the E1 gene region, normally 293 cells (Graham and Prevec, 1991). Ad $E1^-$ recombinants propagated on 293 cells can then be used to 'infect/transfect' a target cell population, where efficient expression of a transgene under the control of a constitutive promoter can be obtained.

For gene therapy, transgenes are preferably inserted in place of the deleted E1 gene region, although other sites on the viral genome can be used. Ad can package only 1.5 kb of sequence in addition to the full-length viral genome, but deletion of both E1 and E3 can permit inserts of up to 7.5 kb to be accommodated. The E3 gene region is dispensable for growth *in vitro* but is known to encode functions that down-regulate cellular MHC class I presentation and render infected cells resistant to tumour necrosis factor and so may have a role in evading immune surveillance *in vivo*. An efficient elegant technology has been developed for the generation of recombinants, based on routine cloning and DNA transfection techniques. The two methods most commonly used provide positive selection for the generation of recombinant virus and thus avoid the need to screen large numbers of viral plaques. The transgene, under the control of the appropriate promoter, is first cloned between flanking Ad sequences in the transfer vector (*Figure 5a*). In the first system, purified viral DNA is cleaved with a restriction enzyme near the E1 gene region, which destroys its ability to replicate when transfected into cells. In the alternative strategy, a plasmid carrying the entire Ad5 genome (pJM17) is used, pJM17 has been constructed so that the size of the prokaryotic vector (pBRX)

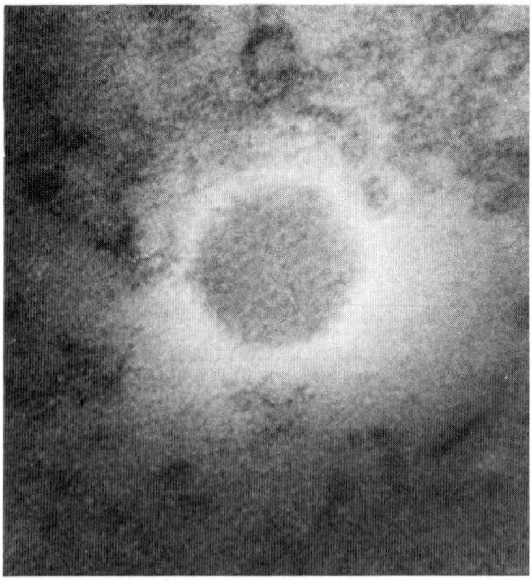

Figure 3. Electron micrograph of an icosohedral Ad particle (~80 nm diameter) recognized in a clinical specimen. Courtesy of Mr Malcolm Smith, Cardiff PHL.

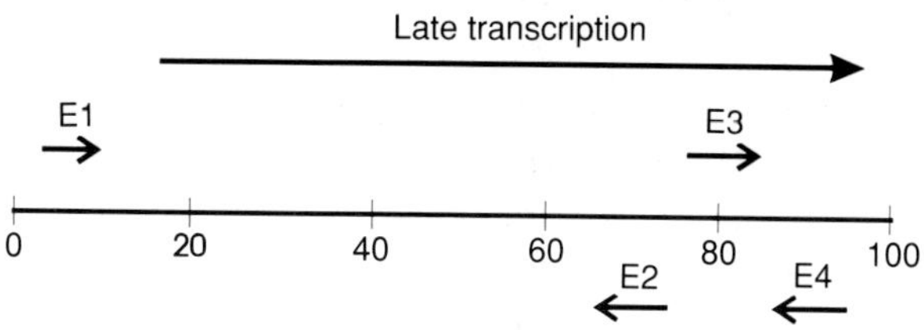

Figure 4. Simplified transcriptional map of the 36 kb adenovirus genome, which is conventionally divided into 100 map units. Ad has a cascade system of gene regulation in which the E1 gene region is transcribed first and activates transcription of all early gene regions. Late transcription, driven by the major late promoter, occurs following replication of the virus genome. Ad transcription is exceptionally complex, due to multiple and alternate splicing of both early and late pre-mRNAs.

inserted in the E1 gene makes the construct too large to package into the virus particle. Recombinant virus is generated by homologous recombination between the transfer vector and either purified genomic viral DNA or pJM17 following transfection of 293 cells (*Figure 5b*). With the first approach, recombination will repair the deletion (generated by the restriction enzyme) at the end of the genome, whilst the second recombination will reduce the size of the viral genome below the packaging limit of the Ad particle; in both cases only the recombinant virus will be capable of generating plaques (Graham and Prevec, 1991). Vectors based on Ad types 5 and 2 are most commonly used, although recombinants have also been generated using the vaccine strains Ad4 and Ad7.

3.2. Gene delivery

Adenoviruses have historically been used as a model system by molecular biologists to study cellular processes and, consequently, knowledge of how these viruses 'work' is relatively advanced. Adenoviruses are also comparatively 'user-friendly'. The replicative cycle can be completed in less than 30 h, up to 10 000 infectious particles can be produced by a single cell and virus can be harvested easily from infected cells. Stocks with titres greater than 10^9 p.f.u. ml^{-1} can readily be produced and virus can be further concentrated and purified by CsCl gradient centrifugation. This facility to generate concentrated stocks is an important feature for a non-replicating vector, as large amounts of virus may be necessary to ensure efficient gene transfer *in vivo*. Recombinant adenoviruses can readily express a transgene in 100% of

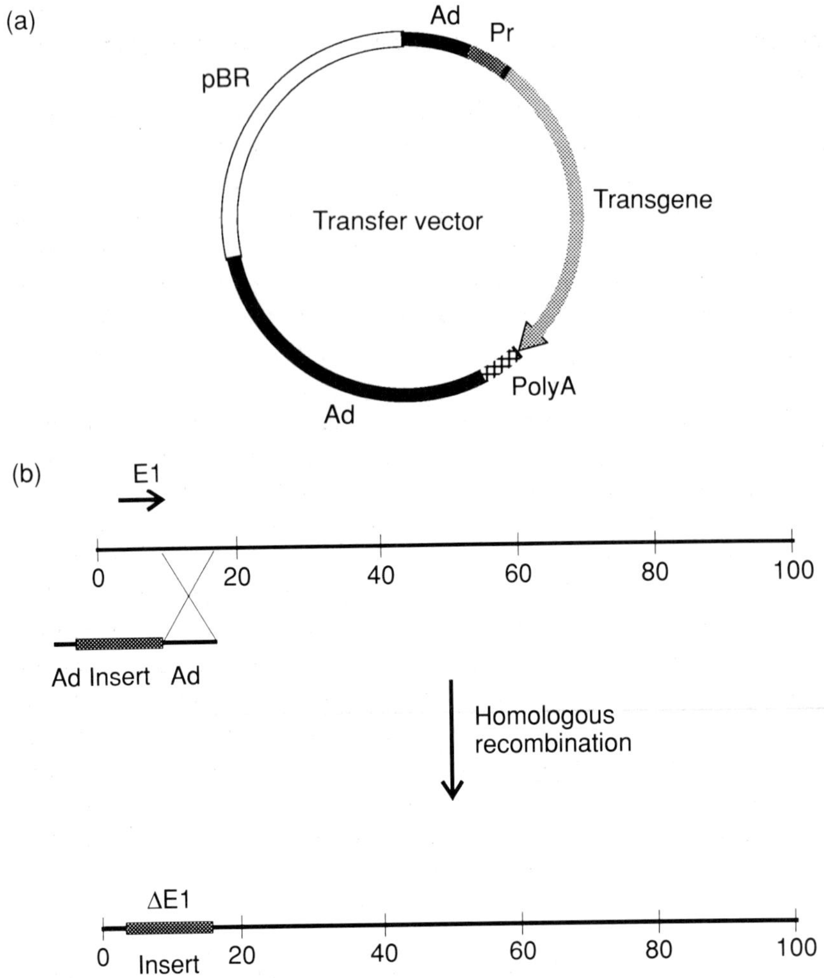

Figure 5. (a) As a first step in the construction of Ad recombinants, genes under the control of an appropriate promoter are normally cloned into a transfer vector such as pXCX2 (Graham and Prevec, 1991) as shown here. (b) The transfer vector is co-transfected into 293 cells with either purified Ad DNA or a plasmid carrying the Ad genome such as pJM17 (Graham and Prevec, 1991) and, following homologous recombination, the transgene is inserted in place of E1 to generate a recombinant virus that will replicate in the helper cell line.

cells *in vitro*, while a number of studies have demonstrated expression from Ad encoding β-galactosidase in an extremely high proportion of targeted cells *in vivo* (Simon *et al.*, 1993, and references therein). Stocks of replication-defective virus are titred in permissive (293) cells but the infectability of other cell types is usually lower and relatively high 'multiplicities of infection' can be required to ensure transgene expression.

Ad E1⁻ vectors intrinsically provide only transient expression, which can be perceived as being either a weakness or a strength, depending on the application. As yet, no specific mechanism to maintain or replicate the viral genome in non-permissive cells has been introduced into the vector. Nevertheless, surprisingly long-term expression has been achieved and this has greatly enhanced the potential range of gene therapy applications. In fibroblasts infected in tissue culture with a replication-defective Ad recombinant, expression continued to increase over time, which suggests that the viral vector may help to maintain the transgene in a transcriptionally active state (Wilkinson and Akrigg, 1992). *In vivo* expression from Ad $E1^-$ vectors has been detected for over 6 months in neurons (Neve, 1993) and muscle cells (Vincent *et al.*, 1993), although, in other systems, transgene expression has been less durable (e.g. expression from AdCFTR and AdLacZ constructs in the respiratory lining of baboons was significantly reduced within 3 weeks; Simon *et al.*, 1993). A variety of factors can be expected to dictate the duration of expression, including rate of target cell turnover, degradation of the vector genome, an immune response to the expressed transgene or down-regulation of the promoter-driving expression. The potential does exist to incorporate elements into the Ad genome to promote long-term maintenance (e.g. an active origin of replication) or chromosomal integration of the transgenes, but such elements would alter the basic properties of the vector and would have to be used with caution.

The ability to express a recombinant gene product is dependent on a combination of the gene delivery system (Ad) and the promoter driving expression. Ad5 is a particularly efficient vehicle for gene delivery in that it will infect (although not necessarily replicate in) most cell types from a wide range of animal species. Ad-mediated gene delivery to different cell targets can be assessed in animal model systems. Transgene expression has been detected in respiratory endothelial cells, the endothelium of arteries and veins, lymphocytes, macrophages, hepatocytes, neurons, glial cells and astrocytes (reviewed in Wilkinson, 1994). *Figure 6* illustrates rat glial cells, infected with AdLacZ (RAd35), expressing β-galactosidase. In contrast to retroviral systems, Ad recombinants are able to promote transgene expression following infection of non-dividing and terminally differentiated cells. In developing the system, strong promoters active in a wide range of cell types have been chosen, such as the Rous sarcoma virus (RSV) LTR or cytomegalovirus

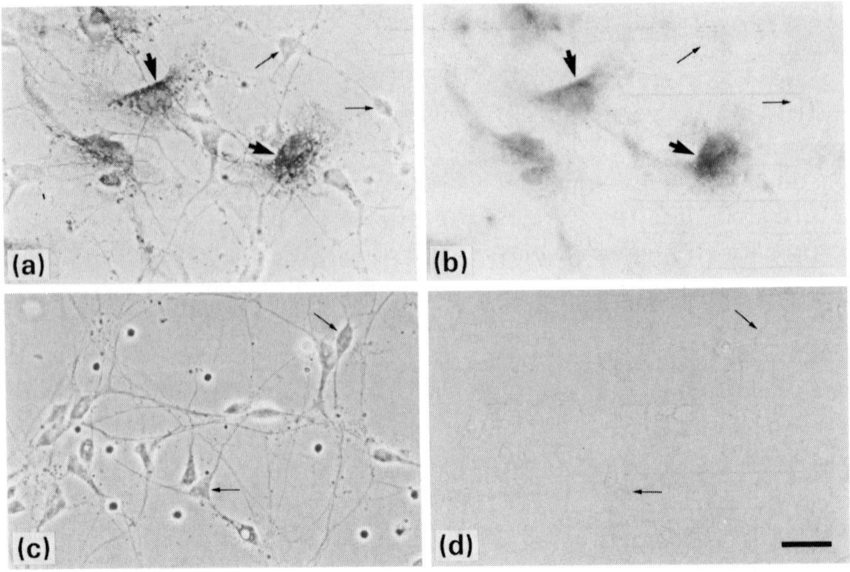

Figure 6. Primary cultures of rat neocortex containing both neurons and glia were infected with a replication-deficient Ad recombinant (RAd35; Wilkinson and Akrigg, 1992) and stained 3 days post-infection for expression of β-galactosidase; positive cells stained blue with the chromogenic substrate X-gal. Cells are shown both under phase contrast, to illustrate cell morphology (a and c), and brightfield, to illustrate the histochemical assessment of β-galactosidase activity (b and d). (a) and (b) are RAd35-infected while (c) and (d) are uninfected. Notice with this particular recombinant that non-neuronal cells (short thick arrows in a and b) display β-galactosidase activity, while neuronal cells (indicated by long fine arrows in a and b) are devoid of such activity. No enzyme activity was detected in the absence of RAd35 infection. Scale bar (shown in d) = 40 μm.

(CMV) major IE promoter. As can be seen from *Figure 7*, replication-defective Ad recombinants are capable of inducing the synthesis of large amounts of recombinant protein, which may be especially useful for producing secreted gene products. Expression levels may be controlled by gene dosage (i.e. the amount of Ad recombinant in the inoculum) and by using a promoter of appropriate strength. Furthermore, the aspiration that targeted expression may be achieved by utilizing cell-specific promoters to drive transcription is supported by studies of gene regulation using Ad recombinants (Ruether *et al.*, 1986).

3.3. Limitations and vector development

Concern has been expressed about the potential toxicity which may be associated with the use of Ad recombinants, especially when multiple

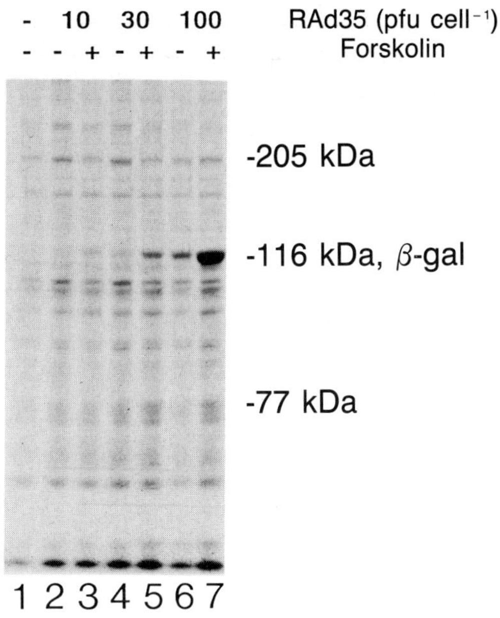

Figure 7. High level expression from the replication-deficient Ad recombinant RAd35 in primary human fibroblasts where 35% of total cell protein can be β-galactosidase when the gene is placed under the control of the powerful constitutive CMV major IE promoter. Levels of transgene expression can be controlled by altering the multiplicity of infection (p.f.u. cell^{-1}) and were enhanced when an inducer of the CMV promoter (10 mM forskolin) was added to cells. The titre of RAd35 was estimated on 293 cells, the infectivity on human fibroblasts was approximately 10-fold lower. Reproduced from Wilkinson and Akrigg (1992) with permission from Oxford University Press.

treatments are required to maintain a therapeutic effect. A component of the virus particle, the penton base protein, can be cytotoxic at high concentration, and extremely high multiplicities of infection can result in limited virus replication. Additionally, certain cell types possess an endogenous E1a-like activity which could potentially enable limited virus replication to occur. The use of AdCFTR recombinants in the treatment of cystic fibrosis patients has highlighted both the strengths and weaknesses of the vector (Engelhardt *et al.*, 1993; Simon *et al.*, 1993). Preliminary experiments have shown efficient expression of either LacZ or CFTR when delivered to the respiratory epithelium of a number of animals and in restricted human trials. The most detailed published study, based on Ad-mediated gene transfer to baboons, has demonstrated that high dosages can induce an inflammatory response in the lung that is believed to be induced by Ad protein in the inoculum. These results are pertinent to the human gene therapy trials which have been initiated but, at the time of writing, have still to be evaluated. It may be possible to hit a window of virus dosage that allows sufficient transgene expression to produce a therapeutic effect without producing a toxic response in human subjects. Furthermore, a specific immune response to Ad structural proteins can also be generated which, in at least one murine model system, has been shown to prevent transgene expression on subsequent treatment with a recombinant (Smith *et al.*, 1993).

As a highly efficient transient expression system, Ad recombinants would appear to be most suited to short-term therapies, such as those which may be involved in wound repair, cell tagging, cancer treatments or vaccination, rather than long-term treatment of a genetic defect. The potency of Ad vectors in obtaining *in vivo* gene transfer is, however, the driving force behind its application for treating diseases such as cystic fibrosis. The results of current gene therapy trials are highlighting areas in which Ad vectors may be improved by selecting an Ad strain with low toxicity, eliminating the potential for rescue with wild-type virus, enhancing infectivity, providing a mechanism for sustained expression and permitting the cloning of larger inserts. Ad has only relatively recently been adopted as a vector for gene therapy and most of the ongoing research is still aimed at evaluating the full potential of the system. Considerable scope exists to tailor Ad vectors to meet the requirements of gene therapy protocols.

4. Adeno-associated virus (AAV) vectors

The parvoviruses are a group of extremely small viruses (18–26 nm diameter) with a short single-stranded DNA genome of approximately 5 kb. Parvovirus B19, the only member of the family to be definitively associated with human disease (erythema infectiosum and aplastic crisis), is capable of autonomous replication. AAVs (genus: *Dependovirus*), however, are unusual in that they require co-infection by either an Ad or herpesvirus in order to replicate. There is intense interest in AAV because, like retroviruses, they integrate and are stably maintained in genomic DNA of infected cells and subsequently are transmitted to daughter cells, following cell division. AAV as a vector, however, benefits from integrating at a specific site in chromosome 19, which can be expected to reduce the potential for insertional mutagenesis and provide a more consistent level of transgene expression. Expression of the AAV *Rep* gene may, however, be necessary for targeting specific provirus integration in the target cell. AAV vectors may be able to overcome other limitations of retroviral systems, including low titre virus stocks, lability of virus in storage and virus inactivation by complement, while permitting transgene expression in non-dividing cells and being able to transduce a wide range of cell types efficiently.

The widespread adoption of AAV vectors has been delayed because of technical problems associated with their use. All protein coding capacity is normally deleted from AAV vectors, providing room to clone a transgene of up to 5 kb. Recombinants therefore require functions to be provided in *trans*, not only for AAV-encoded genes but also for the helper adenovirus (Muzyczka, 1992). Difficulties associated with providing all these helper functions simultaneously has made the production of high titre stocks of recombinant AAVs problematical; however, this obstacle should be surmountable. The cloning of certain transgenes into the single-stranded DNA vector has also been incompatible with vector replication. Nevertheless, the potential advantages of the system merit perseverance: recent encouraging studies have demonstrated expression of CFTR in the lung of a rabbit for up to 6 months following *in vivo* application (Flotte *et al.*, 1993) and a recombinant has been produced encoding HIV anti-sense sequences that has been shown to inhibit HIV replication *in vitro* (Chatterjee *et al.*, 1992).

5. Herpesvirus vector systems

Direct gene delivery into post-mitotic neurons *in vitro* (*Figure 8*) and *in vivo*, through the use of viral vectors, has accelerated the development of gene therapy for neurological disorders (Friedman and Jinnah, 1993; Wolfe *et al.*, 1992). The two herpesviruses most commonly used in these studies, herpes simplex virus type 1 (HSV-1) and pseudorabies virus, are both neurotropic and able to establish latency in neurons (Enquist, 1994; Roizman and Sears, 1991). Herpesvirus vectors have been exploited in recent years to dissect and manipulate neuronal physiology at the molecular level (Geller *et al.*, 1991). Furthermore, the detailed understanding of the specific trans-synaptic interneuronal spread of pseudorabies and HSV-1 during infections of the central nervous system (CNS) has produced a powerful and very elegant tool to examine neuronal pathways of synaptically linked networks (Enquist, 1994). Potentially, this technology could be used to deliver genes to all constituent neurons within a given distributed neuronal network,

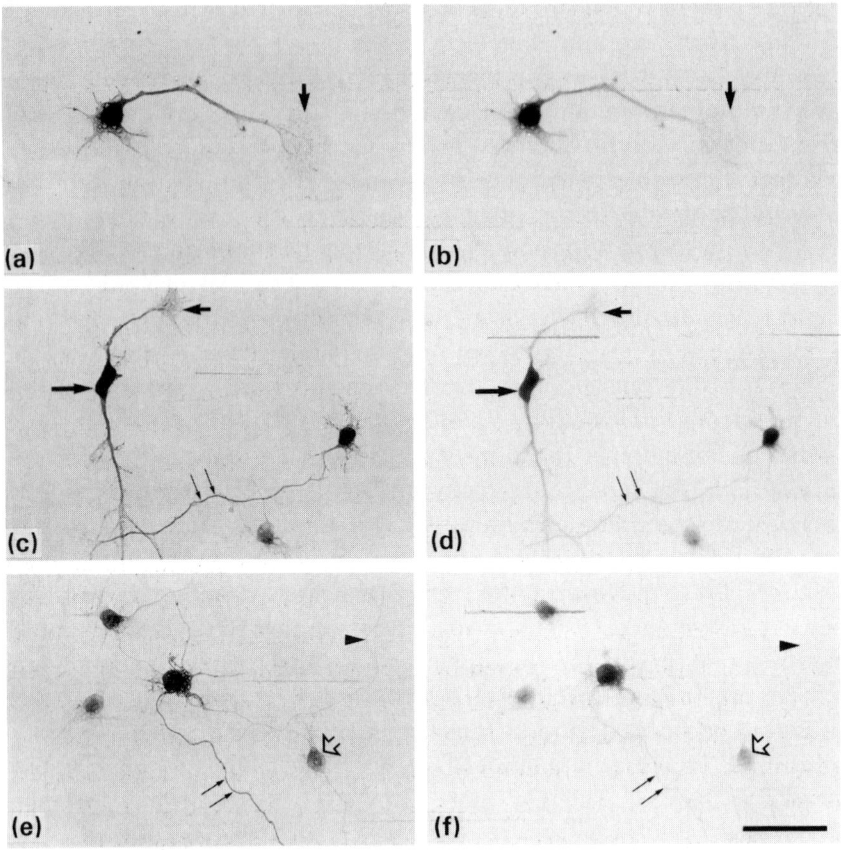

Figure 8. Expression of chloramphenicol acetyl-transferase in neocortical neuronal cultures infected with HSV-1 *ts*K/CAT (CAT under transcriptional control of the HSV-1 IE3 promoter in the thymidine kinase locus of the HSV1 mutant *ts*K; Lowenstein *et al.*, 1994) 18 h post-infection. Morphological characteristics of neurons are seen under phase contrast (a, c, e) while the same fields under brightfield illumination illustrate immunoreaction with a mouse monoclonal antibody to CAT (b, d, f). (a) and (b) show a neuron expressing CAT, where the enzyme has diffused into the axonal growth cone (arrow). In (c) and (d) immunoreactivity has diffused throughout all neuronal processes, including a growth cone (short thick black arrow) and along an axon (fine black arrows). In (d) notice also the gradient in enzyme immunoreactivity diffusing from the nucleus of the neuron to the left (long thick black arrow) into all the cell's neurites. (e) and (f) illustrate different levels of transgene expression. The axon indicated by a pair of long fine black arrows is immunoreactive for CAT all along its length, while in the neuron whose cell body is indicated by an open arrow, only low level of immunoreactivity within the nucleus, but not the axon (black arrowhead), was detected. In the absence of viral infection, no cells gave positive staining (results not shown). Scale bar (shown in f) = 40 μm.

rather than just to restricted areas of the brain.

5.1. Herpesvirus replication cycle and its relevance to herpesvirus vectors

Productive or lytic cycle. HSV-1 has a large double-stranded DNA genome of 153 kb which has been sequenced and shown to encode approximately 73 genes. The viral particle (virion) consists of a central core containing the HSV-1 genome surrounded by a regular icosahedral capsid, a region of tegument and, finally, the envelope. Following attachment to the cell surface, the virion envelope fuses with the plasma membrane and the nucleocapsid (core/capsid), and tegument proteins enter the cell. The nucleocapsid then transports the viral genome to the nucleus. HSV-1 gene expression is conventionally divided into three phases of viral gene expression, α (immediate–early (IE)), β (early) and γ (late) (reviewed in Roizman and Sears, 1991). There are only five IE genes which, by definition, are transcribed in the absence of *de novo* virus-encoded protein synthesis. IE transcription is strongly stimulated by a component of the virion tegument, Vmw65 (also called VP16 or αTIF), which plays an important role in activating the productive cycle of HSV-1 infection. During a productive infection, viral gene expression proceeds through all three phases, leading to the production of new viral particles. Virion maturation comprises a series of complex steps which involve the budding of the virus through the inner nuclear membrane, transport through the cell's secretory apparatus (ER and Golgi), and release of the viral particle through the plasma membrane (Roizman and Sears, 1991).

During natural infection of sensory neurons, the viral particles are taken up by the axonal terminals and transported to the cell body. If replication occurs, new virus particles will be transported to dendrites, from where they will be released at synaptic sites into the intersynaptic space, cross synaptic junctions and enter the next neuron in a synaptically linked network. Under experimental conditions, the virus can enter neurons through neuronal surfaces other than axon terminals. The herpesvirus nucleocapsid is capable of being transported in either antero- or retrograde fashion inside axons (Enquist, 1994; Roizman and Sears, 1991).

Latent cycle. HSV-1 can establish latency upon entering neurons. Latency is defined as the permanence of the virus genome in cells, without any overt manifestations of ongoing viral infection. The latency cycle of HSV can be divided into three different stages: establishment, maintenance and reactivation. The molecular mechanisms leading to the establishment of latency are not completely understood, although two rival hypotheses are currently being examined. Briefly, the first states that no HSV-1 gene expression occurs during the entry of HSV-1 into latency while, in the second, limited low level viral gene expression and replication may occur but without virus production. Understanding latency has important implications for vector design aimed at long-term transgene expression (Enquist, 1994; Steiner and Kennedy, 1993). In order to encourage HSV-1 vectors to establish and maintain latency (i.e. not reactivate), vectors based on defined genetic defects in immediate–early or early genes are being generated (Breakfield *et al.*, 1992; Glorioso *et al.*, 1992). A cellular transcription factor has recently been identified as an inhibitor of HSV-1 viral transcription (Oct-2 neuronal-specific isoforms) that may have a role in maintaining HSV-1 latency (see Chapter 13); there may be potential to exploit this model of transcriptional repression to promote latency in HSV-1 vectors.

Once latency is established, only one portion of the HSV-1 genome is transcriptionally active. This region expresses a small number of RNAs, called latency-associated transcripts (LATs). Two LAT promoters and two potential open reading frames have been identified in this region (Glorioso *et al.*, 1992), although whether the putative proteins are expressed remains uncertain (Steiner and Kennedy, 1993). LAT promoters have been used to drive expression in HSV-1 recombinants (Wolfe *et al.*, 1992). Long-term expression is dependent on location of transgene insertion into the HSV-1 genome and the target cell type. Most encouraging results have been obtained using a combined LAT/MoMLV promoter, with expression of a reporter gene in murine dorsal root ganglion and trigeminal ganglion neurons for up to 6 months having been achieved (Steiner and Kennedy, 1993). While expression from HSV-1 recombinants in dorsal root ganglion, trigeminal ganglion and brain stem neurons can be stable, possibly permanent, expression in cells of the CNS is still, at best, only transient (1–2 months) (Breakfield *et al.*, 1992; Glorioso *et al.*, 1992; Steiner and Kennedy, 1993; Wolfe *et al.*, 1992).Whether latency can be established in other neurons located throughout the CNS is currently hotly debated.

5.2. Herpesvirus vector systems

Two quite different types of HSV-1 vectors have been developed: in conventional HSV-1 recombinants the gene of interest is cloned directly into the viral genome, while in HSV-1 amplicons the gene of interest is cloned into a subgenomic replicating element (see *Figure 9*).

Recombinant HSV-1 vectors. Usually genes are inserted into the HSV genome by homologous recombination, although methods have been developed based on ligation into unique restriction enzyme sites within the HSV-1 genome and the use of the Cre-lox recombination system from coliphage P1 (Breakfield *et al.*, 1992; Glorioso *et al.*, 1992; Leib and Olivo, 1993; Lowenstein *et al.*, 1994; Wolfe *et al.*, 1992). A potential advantage of HSV-1 as a vector is that some 38 genes are non-essential for virus growth in tissue culture cells and thus, theoretically, these sequences could be deleted, freeing up space for the insertion of 30–35 kbp of foreign DNA. Such vectors would permit insertion of long spliced cellular genes, rather than cDNA copies, and the use of long promoter/regulatory elements to drive transcription. HSV-1 vectors permit the use of cell-specific promoters and, indeed, neuronal-specific promoters have provided targeted expression (Breakfield *et al.*, 1992). In addition, a LAT promoter has been shown to function in a neuron-specific manner (Steiner and Kennedy, 1993).

Disabled HSV-1 vectors based on mutated viruses are routinely used, including: temperature-sensitive mutants (e.g. HSV*ts*K which has a mutant IE gene), mutations in genes required for growth in some cell types (e.g. thymidine kinase deficient) and deletion mutants for genes essential for growth, which have to be grown on a *trans*-complementing cell line (Breakfield *et al.*, 1992; Glorioso *et al.*, 1992; Lowenstein *et al.*, 1994; Wolfe *et al.*, 1992). Any of these vectors, will be unable or severely reduced in its capacity to complete its replicative cycle in neurons; but nevertheless can establish latency.

Amplicon vectors. The generation of recombinant amplicons is rapid, since the gene of interest is inserted directly into a bacterial plasmid containing the relevant HSV sequences (a replication origin and 'a' sequence). This plasmid is then either co-transfected into cells together with helper virus DNA, or transfected into cells which are then infected with helper virus. Functions provided by the helper virus will replicate and package the amplicon into HSV-1 virions, the origin permits amplicon replication and the 'a' sequence is required for packaging amplicon DNA into the HSV-1 nucleocapsid (*Figure 9*). During HSV-1 replication viral DNA consists of head-to-tail concatemers, which suggests that replication occurs by a rolling circle mechanism (Roizman and Sears, 1991). The helper virus packages a single 153 kb copy of its genome, while the amplicon will contain multiple copies, up to the upper packaging limit of HSV virions (Geller *et al.*, 1991; Leib and

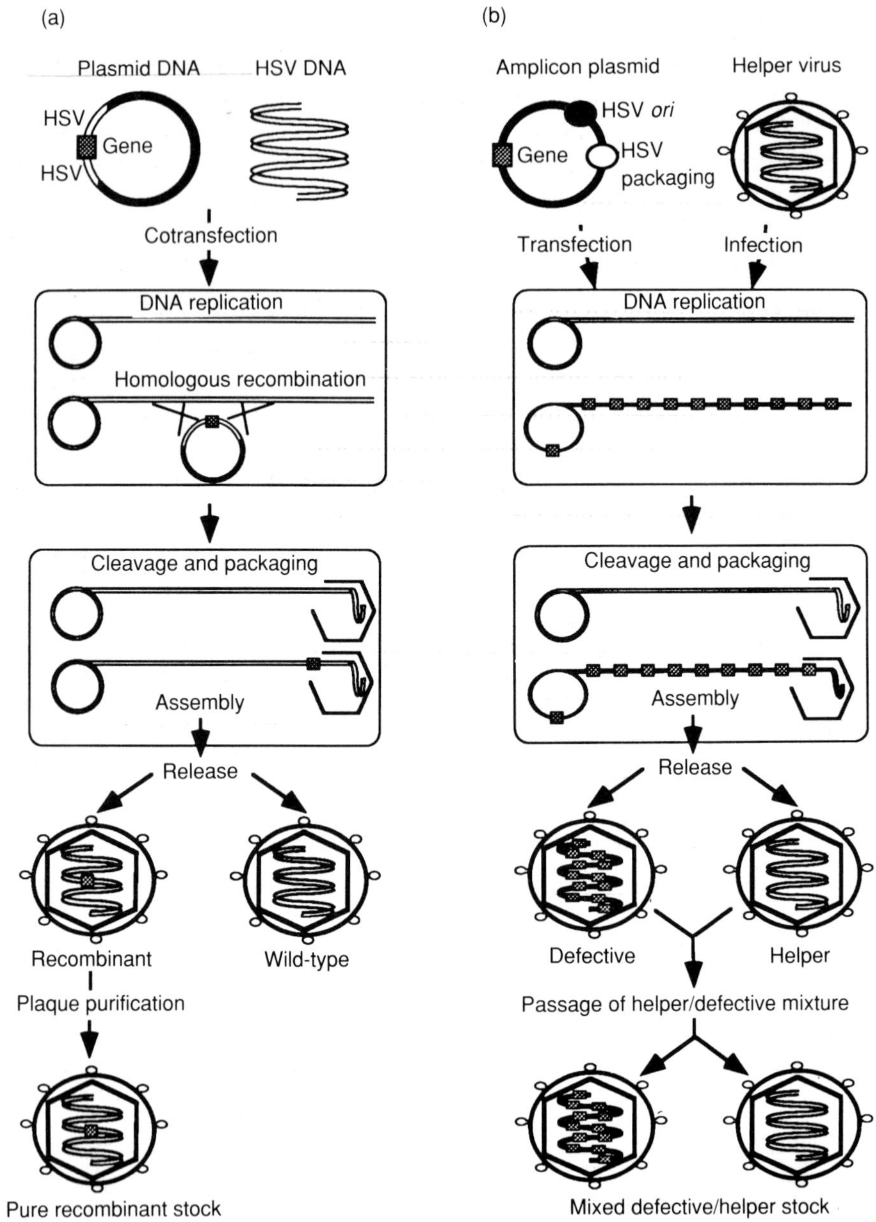
(a)
Plasmid DNA
HSV DNA
HSV
Gene
HSV
Cotransfection
DNA replication
Homologous recombination
Cleavage and packaging
Assembly
Release
Recombinant
Wild-type
Plaque purification
Pure recombinant stock
(b)
Amplicon plasmid
Helper virus
HSV ori
Gene
HSV packaging
Transfection
Infection
DNA replication
Cleavage and packaging
Assembly
Release
Defective
Helper
Passage of helper/defective mixture
Mixed defective/helper stock

Olivo, 1993). There are few limitations in the choice of promoter or DNA sequence to be cloned into HSV amplicons (Geller *et al.*, 1991; Leib and Olivo, 1993). Amplicons are clearly replication-defective vectors, however, any amplicon preparation will contain a proportion of helper virus. The amplicon/helper virus ratio can be variable and is awkward to determine, while the replication competence of the HSV-1 helper virus is also important (Leib and Olivo, 1993).

5.3. Vector toxicity

The general cytotoxicity of HSV-1 vectors constitutes an important limitation of current systems. Toxicity is associated with: (i) virion proteins, such as the virus host shut-off function, and (ii) certain gene products expressed during the lytic cycle which are toxic to cells (e.g. thymidine kinase, ribonucleotide reductase, IE gene products). Even non-replicating vectors deficient in the production of an essential IE protein can be cytotoxic when used to infect at high multiplicities of infection; the mechanism of this phenomenon is still being studied.

Figure 9. A schematic representation of the two most common methods for generating HSV vectors. (a) Recombinant HSV vectors. Genomic viral DNA from any wild-type or mutant HSV is isolated and transfected into permissive cells along with a plasmid containing a foreign gene (▩) inserted into a fragment of the HSV genome (□). Transfection of viral DNA initiates the virus lytic cycle and viral DNA replication. The HSV DNA sequences on the plasmid DNA undergo homologous recombination with viral DNA. Replicated and recombined viral DNA is packaged into new virus particles. Progeny virus is predominantly of the parental type, with a low percentage of recombinant virus containing the foreign gene. Recombinant virus is then plaque-purified away from parental virus. (b) Defective virus derived from an HSV amplicon plasmid. In addition to sequences for propagation in bacteria, an amplicon plasmid contains an HSV origin of replication (⬬) and a cleavage-packaging signal (⬭). The foreign gene (▩) is cloned into the amplicon plasmid, which is transfected into cells permissive for the helper virus. The transfected cells are then infected with a helper virus which provides all the regulatory and structural genes necessary for viral growth. The helper virus can be a mutant with a lesion in an essential gene, in which case a complementing cell line is used. The helper virus provides the functions for rolling circle replication of both the helper virus DNA and the amplicon DNA. Helper virus DNA and a linear head-to-tail concatemeric form of the amplicon plasmid, containing multiple copies of the foreign gene, are packaged separately into newly made virus particles. Both defective and helper virus particles are released from the cell. The defective virus can infect cells but cannot replicate in the absence of helper virus. Also defective and helper virus cannot be physically separated. This figure and modified legend reproduced from Leib and Olivo (1993) with kind permission from Drs D. Leib and P. Olivo and ICSU Press.

Several research groups are trying to overcome these problems by generating combined mutants lacking several of the IE genes, or creating mutations in further essential genes (Breakfield *et al.*, 1992; Friedmann and Jinnah, 1993; Glorioso *et al.*, 1992). Since amplicons are packaged in HSV-1 virions and are dependent on helper virus, they do not by themselves overcome the problem of toxicity (Breakfield *et al.*, 1992; Geller *et al.*, 1991; Glorioso *et al.*, 1992; Leib and Olivo, 1993).

5.4. Possible applications for gene therapy

HSV vectors have opened up the possibility of developing gene transfer vectors for neuronal delivery, with the prospects of offering new treatments for a great number of neurological disorders, both sporadic and inherited (Friedman and Jinnah, 1993). Currently, developments are focussing on the production of long-term, high-level, non-toxic, expression vectors. While HSV vectors containing genes of interest for neurobiological and neuro-therapeutic applications have already been produced, they have not yet been applied in animal models of human neurological disease (Friedmann and Jinnah, 1993). Nevertheless, the available vectors are proving to be invaluable in the exploration of neuronal structure, physiology, cell biology and basic mechanisms of cell death (Geller *et al.*, 1991; Lowenstein *et al.*, 1994).

HSV vectors can be used for short-term expression in non-dividing cells (neurons), very slowly dividing cells (hepatocytes) and other cell types. Additionally, in dorsal root ganglion and trigeminal ganglion neurons, HSV-1 recombinants have been used to stably, and possibly permanently, transduce transgene products. This is of particular importance, since a number of inherited neurological disorders specifically affect these neurons (e.g. inherited ataxias and peripheral neuropathies). Within this context, it is important to emphasize the shortcomings of current neurological therapeutics. Effective treatments are limited to a handful of neurological disorders. The majority of such treatments are purely symptomatic, without any effect on disease progression and long-term outcome. HSV and, more recently, Ad, vectors have opened up new avenues towards developing effective strategies which aim to alter the genetic causes, or modify the physiopathological basis, of inherited and sporadic neurological diseases. Nevertheless, the general toxicity of HSV vectors, the possibility of reactivation of and recombination with latent wild-type HSV, and the possibility of herpes encephalitis, are all aspects which will have to be addressed before these experimental techniques can be applied in the treatment of human disease.

6. When is a viral vector not a viral vector?

Perhaps the greatest concern with using recombinant viruses in gene therapy is that co-infection with a wild-type virus could 'rescue' and thereby produce lateral spread of even replication-defective viral vectors. The packaging constraints of the virus particles often restrict the size of transgene insertion, and toxicity or a risk of cellular transformation can accompany the use of some systems. There is clearly scope for direct DNA transfer systems, such as lipofection and particle bombardment, to provide an alternative or complementary strategy for gene delivery. Gene therapy must continue to exploit the best available technology to optimize gene delivery. Experience with viral vectors is, however, very relevant to direct DNA transfer technology. Viruses provide a remarkable reservoir of characterized genetic elements designed to deliver and regulate expression of foreign genes in cells. The strong constitutive CMV major IE and RSV LTR promoters are widely used to ensure high level expression, the affinity of certain virion proteins for cell-specific receptors may be utilized to target expression (e.g. in lipofection), and the *tr* elements which flank the AAV genome could potentially be exploited to direct transgene insertion into chromosomal DNA in virus-free systems. Episomes are perhaps the most widely used vectors based on a viral genetic element and they may also prove to be of value in promoting stable expression in gene therapy (Margolskee, 1992).

Episomal vectors are plasmids which will replicate and be maintained as extrachromosomal elements at low copy number (~5–20 molecules per cell) and are stably transmitted to daughter cells, although the term episome can also be used to describe any genetic element which will take up a circular DNA conformation (e.g. the HSV genome following infection). Systems have been developed based on the Epstein–Barr virus (EBV), BK virus and bovine papillomavirus origins of replication. Since episomal vectors are designed to replicate in both prokaryotic and eukaryotic cells, they are relatively easy to manipulate and they have been used extensively to express proteins in continuous cell lines. High level expression is associated with maintaining genes on an extra-chromosomal element. An EBV-based episome has been used in conjunction with lipofection to deliver the human CFTR gene in a murine system (Hyde *et al*, 1993). However, perhaps the best demonstration of the potential of such systems to gene therapy comes from the effective way that the viruses themselves can sustain life-long infections.

Ad particles enter cells by means of receptor-mediated endocytosis

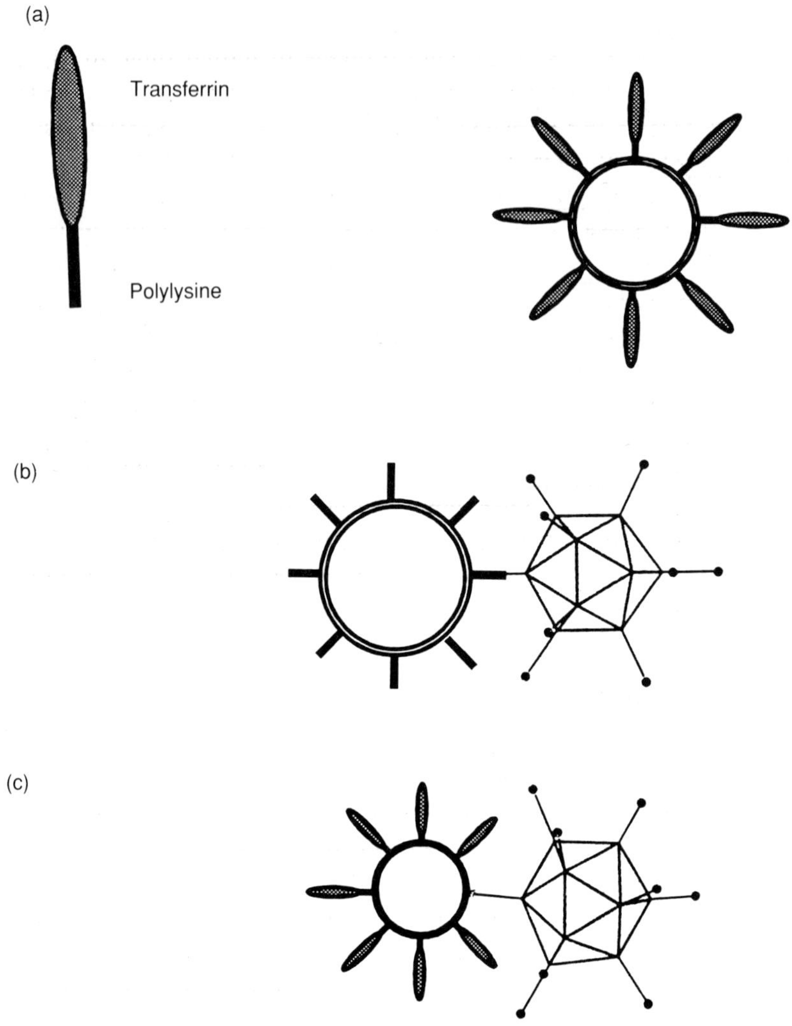

Figure 10. (a) In receptor-mediated endocytosis, DNA is coupled by means of a polycation (polylysine shown here) bridge to a specific ligand (in this case transferrin). Transferrin will bind its receptor on the cell surface to promote DNA uptake. (b) Ad–polylysine–DNA complexes provide enhanced expression following receptor-mediated endocytosis. The Ad particle has a specific mechanism by which it is released from endosomes and this is also thought to help uptake of the complexed DNA. (c) Yet more efficient expression is obtained when the complex includes transferrin, which has high affinity for its receptor on cell targets and promotes efficient endocytosis.

and contain a specific mechanism associated with the penton base protein to release the virus particle into the cytosol as the pH in the endosome drops. Birnsteil and colleagues (Zenke *et al.*, 1990) have developed a novel approach to gene transfer in which they attached DNA to the *outside* of an Ad particle (*Figure 10*). The Ad particle binds to the cell and escorts the transgene through the endocytic vesicle, this appears to protect the DNA from being degraded, and the process produces significantly higher levels of expression than conventional receptor-mediated endocytosis. In a further development of the system, the transferrin receptor was combined with the Ad particle:DNA complex (Wagner *et al.*, 1992). Using transferrin to direct binding of the complex to the target cell substantially increases the efficiency of transgene delivery. When a chicken Ad was substituted for the human Ad in the latter system the cytotoxicity associated with Ad proteins was much reduced (Cotton *et al.*, 1993). Ad particles are able, impressively, to provide expression of a reporter gene in nearly 100% of cells, and the system has been adapted, with some success, to *in vivo* gene transfer (Gao *et al.*, 1993). Ad particles are an impressive example of how viral components may be exploited and adapted to generate 'virus-free' gene delivery system.

Acknowledgements

We apologize to the many researchers concerned that the citation of original papers in this review has been limited by space. P.R.L. is a Research Fellow of The Lister Institute of Preventive Medicine and wishes to acknowledge the support from SERC, The Wellcome Trust, MRC, The Royal Society, and The Welsh Scheme for the Development of Health and Social Research. We wish to thank Professor David Lane, for his kind supply of monoclonal antibodies against CAT, and Professor Lynn Enquist's constructive comments on the herpes virus vector section.

References

Aebersold P, Kasid A and Rosenberg SA. (1990) Selection of gene-marked tumour infiltrating lymphocytes from post-treatment biopsies: a case study. *Hum. Gene Ther.* **1**, 373–384.

Anderson WF. (1993) Editorial: What about those monkeys that got T-cell lymphoma? *Hum. Gene Ther.* **4**, 1–2.

Beck-Engeser G, Stocking C, Just U, Albritton L, Dexter M, Spooncer E and Ostertag W. (1991) Retroviral vectors related to the myeloproliferative sarcoma virus allow efficient expression in haematopoietic stem and precursor cell lines, but retroviral infection is reduced in more primitive cells. *Hum. Gene Ther.* **2**, 61–70.

Berkner KL. (1988) Development of adenovirus vectors for the expression of heterologous genes. *BioTechniques,* **6**, 616–629.

Blaese RM. (1993) Development of gene therapy for immunodeficiency: adenosine deaminase deficiency. *Pediatr. Res.* **33**, S49–S55.

Bösze Z, Thiesen H-J and Charnay, P. (1986) A transcriptional enhancer with specificity for erythroid cells is located in the long terminal repeat of the Friend murine leukaemia virus. *EMBO J.* **5**, 1615–1623.

Bowtell DDL, Cory S, Johnson GR and Gonda TJ. (1988) Comparison of expression in haematopoietic cells by retroviral vectors carrying two genes. *J. Virol.* **62**, 2464–2473.

Breakfield XO, Huang Q, Andersen JK *et al.* (1992) Gene transfer into the nervous system using recombinant herpesvirus vectors. In: *Gene Transfer and Therapy in the Nervous System* (eds FH Gage and Y Christen). Springer-Verlag, Berlin, pp. 118–132.

Brenner MK, Rill DR, Moen RC, Krance RA, Mirro J, Anderson WF and Ihle JN. (1993) Gene marking to trace origin of relapse after autologous bone marrow transplantation. *Lancet,* **341**, 85–86.

Burns JC, Friedmann T, Driever W, Burrascano M and Yee, JK. (1993) Vesicular stomatitis virus G glycoprotein pseudotyped retroviral vectors: concentration to very high titre and efficient gene transfer into mammalian and non-mammalian cells. *Proc. Natl Acad. Sci. USA,* **90**, 8033–8037.

Chatterjee S, Johnson PR and Wong KK Jr. (1992) Dual target inhibition of HIV *in vitro* by means of an adeno-associated antisense vector. *Science,* **258**, 1485–1488.

Cornetta K, Moen RC, Culver K *et al.* (1990) Amphotropic murine leukaemia virus is not an acute pathogen for primates. *Hum. Gene Ther.* **1**, 15–30.

Cotton M, Wagner E, Zatloukal K and Birnstiel ML. (1993) Chicken adenovirus (CELO) particles augment receptor mediated DNA delivery to mammalian cells and yield exceptional levels of stable transformants. *J. Virol.* **67**, 3777–3785.

Engelhardt JF, Simon RH, Yang Y, Zepeda M, Weber-Pendleton S, Doranz B, Grossman M and Wilson JF. (1993) Adenovirus-mediated transfer of the CFTR gene to lung of nonhuman primates: biological efficacy study. *Hum. Gene Ther.* **4**, 759–769.

Enquist LW. (1994) Infection of the mammalian nervous system by pseudorabies virus (PRV). *Sem. Virol.,* in press.

Flotte TR, Afione SA, Conrad C, McGrath SA, Solow R, Oka H, Zeitlin PL, Guggino WB and Carter BJ. (1993) Stable *in vivo* expression of the cystic fibrosis transmembrane conductance regulator with an adeno-associated virus vector. *Proc. Natl Acad. Sci. USA,* **90**, 10613–10617.

Friedman T and Jinnah HA. (1993) Gene therapy for disorders of the nervous system. *Trends Biotechnol.* **11**, 192–197.

Gao L, Wagner E, Cotten M *et al.* (1993) Direct *in vivo* gene transfer to airway epithelium employing adenovirus–polylysine–DNA complexes. *Hum. Gene Ther.* **4**, 17–24.

Geller AI, During MJ and Neve RL. (1991) Molecular analysis of neuronal physiology by gene transfer into neurons with herpes simplex virus vectors. *Trends Neurosci.* **14**, 428–432.

Glorioso JC, Goins WF and Fink DJ. (1992) Herpes simplex virus-based vectors. *Sem. Virol.* **3**, 265–276.

Graham FL and Prevec L. (1991) Manipulation of adenovirus vectors. In: *Methods in Molecular Biology, Vol 7, Gene Transfer and Expression Protocols* (ed. EJ Murray). Humana Press, Clifton, NJ, pp. 109–128.

Grunhaus A and Horwitz MS. (1992) Adenoviruses as cloning vectors. *Sem. Virol.* **3**, 237–252.

Horwitz M. (1990) Adenoviruses. In: *Virology* (eds BN Fields and DM Knipe). Raven Press, New York, pp. 1723–1740.

Hyde SC, Gill DR, Higgins CF, Trezise AEO, MacVinish LJ, Cuthbert AW, Ratcliff R, Evans MJ and Colledge WH. (1993) Correction of the ion transport in cystic fibrosis transgeneic mice by gene therapy. *Nature,* **362**, 250–255.

Kempler G, Freitag B, Berwin B, Nanassay O and Barklis E. (1993) Characterisation of the Moloney murine leukaemia virus stem cell-repressor binding site. *Virology,* **193**, 690–699.

Ledley FD, Brody B, Kozinetz CA and Mize SG. (1992) The challenge of follow-up for clinical trials of somatic gene therapy. *Hum. Gene Ther.* **3**, 657–663.

Leib DA and Olivo PD. (1993) Gene delivery to neurons: is herpes simplex virus the right tool for the job? *Bioessays,* **15**, 547–554.

Lowenstein PR, Morrison EE, Bain D, Hodge P, Preston CM, Clissold P, Stow ND, McKee TA and Castro MG. (1994) Use of recombinant vectors derived from herpes simplex virus 1 mutant *ts*K for short term expression of transgenes encoding cytoplasmic and membrane anchored proteins in postmitotic polarized cortical neurons and glial cells *in vitro*. *Neuroscience,* in press.

Margolskee RF. (1992) Epstein–Barr virus based expression vectors. In: *Current Topics in Microbiology and Immunology, Vol. 158, Viral Expression Vectors* (ed. N Muzyczka). Springer-Verlag, Berlin, pp. 67–95.

Miller AD. (1990) Retrovirus packaging cells. *Hum. Gene Ther.* **1**, 5–14.

Miller AD. (1992) Retroviral vectors. In: *Current Topics in Microbiology and Immunology, Vol. 158, Viral Expression Vectors* (ed. N Muzyczka). Springer-Verlag, Berlin, pp. 1–24.

Moolten FL and Cupples LA. (1992) A model for predicting the risk of cancer consequent to retroviral gene therapy. *Hum. Gene Ther.* **3**, 479–486.

Morgan RA and Anderson WF. (1993) Human gene therapy. *Annu. Rev. Biochem.* **62**, 191–217.

Muzyczka N. (1992) Use of adeno-associated virus as a generalised transduction vector for mammalian cells. In: *Current Topics in Microbiology and Immunology, Vol. 158, Viral Expression Vectors* (ed. N Muzyczka). Springer-Verlag, Berlin, pp. 97 129.

Neve RL. (1993) Adenovirus enters the brain. *Trends Neurosci.* **16**, 251–253.

Roizman B and Sears AE. (1991) Herpes simplex viruses and their replication. In: *Fundamental Virology* (eds BN Fields and DM Knipe). Raven Press, New York, pp. 849–895.

Ruether JE, Maderious A, Lavery D, Logan J, Fu SM and Chen-Kiang S. (1986) Cell-type-specific synthesis of murine μ RNA from an adenovirus vector. *Mol. Cell. Biol.* **6**, 123–133.

Simon RH, Engelhardt JF, Yang, Y, Zepeda M, Weber-Pendleton S, Grossman M and Wilson JM. (1993) Adenovirus-mediated transfer of the CFTR gene to lung of nonhuman primates: toxicity study. *Hum. Gene Ther.* **4**, 771–780.

Smith TAG, Mehaffey MG, Kayda DB, Saunders JM, Yei S, Trapnell BC, McClelland A and Kaleko M. (1993) Adenovirus-mediated expression of therapeutic plasma levels of human factor IX in normal mice. *Nature Genetics,* **5**, 397–402.

Steiner I and Kennedy PGE. (1993) Molecular biology of herpes simplex virus type 1 latency in the nervous system. *Mol. Neurobiol.* **7**, 137–159.

Temin HM. (1990) Safety considerations in somatic gene therapy of human disease with retrovirus vectors. *Hum. Gene Ther.* **1**, 111–123.

Tolstoshev P. (1993) Gene therapy, concepts, current trials and future directions. *Annu. Rev. Pharmacol. Toxicol.* **33**, 573–596.

Vincent N, Ragot T, Gilgenkrantz H, Couton D, Chafey P, Gregoire A, Briand P, Kaplan J-C, Kahn A and Perricaudet M. (1993) Long term correction of mouse dystrophic degeneration by adenovirus-mediated transfer of a minidystrophin gene. *Nature Genetics,* **5**, 130–134.

Wagner E. Zatloukal K, Cotten M, Kirlappos H, Mechtler K, Cureil D and Birnstiel ML. (1992) Coupling of adenovirus to transferrin-polylysine/DNA complexes greatly enhances receptor-mediated gene delivery and expression in transfected cells. *Proc. Natl Acad. Sci. USA,* **89**, 6099–6103.

Wilkinson GWG. (1994) Gene therapy and viral vaccination. *Rev. Med. Microbiol.* **5**, 97–106.

Wilkinson GWG and Akrigg A. (1992) Consitutive and enhanced expression from a defective adenovirus vector. *Nucleic Acids Res.* **20**, 2233–2239.

Wolfe JH, Deshmane SL and Fraser NW. (1992) Herpesvirus vector gene transfer and expression of β-glucuronidase in the central nervous system of MPS VII mice. *Nature Genetics,* **1**, 379–384.

Yu M, Poeschla E and Wong-Staal, F. (1994) Progress towards gene therapy for HIV infection. *Gene Ther.* **1**, 13–26.

Zenke M, Steinlein P, Wagner E, Cotten M, Beug H and Birnstiel ML. (1990) Receptor-mediated endocytosis of transferrin–polycation conjugates: an efficient way to introduce DNA into hemopoietic cells. *Proc. Natl Acad. Sci. USA,* **87**, 3655–3659.

11

Direct gene transfer for the treatment of human disease

G.J. Nabel and E.G. Nabel
Howard Hughes Medical Institute, The University of Michigan Medical Center, Ann Arbor, USA

1. Introduction

The impact of molecular genetics in human disease ranges from molecular diagnostics to drug discovery, pathogenesis and genetic treatments. The application of molecular genetics to the treatment of human diseases requires an understanding of the specific molecular defects that must be addressed. It is, therefore, helpful to define the phenotype of human diseases in terms of their precise genetic defect, because the strategy needed to approach them depends upon the underlying problems. The classic examples of genetic diseases are inherited, with mutations in the coding or regulatory sequences of genes responsible for defective proteins, as observed in diseases such as cystic fibrosis, Lesch–Nyhan syndrome and thalassaemia. Acquired diseases are also caused by genetic defects. In a disease like AIDS, for example, foreign genetic material is introduced into human cells. In fact, it is this aspect of AIDS that makes it an attractive target for treatment with gene therapy. Finally, there are diseases where normal genes demonstrate abnormal expression which may induce pathological processes and thus cause disease. The question in each example becomes: how can investigators approach the underlying genetic problems relevant to its pathogenesis.

The basis for much of the work in our laboratories has come from viruses. In the virus life-cycle, there is an afferent phase of infection, in which genetic material is brought into the nucleus, and an efferent phase in which viral genes are expressed. These studies focus on the regulation of latency. In HIV, for example, what regulates the packaging and the export of viral genes from the nucleus and into a mature particle? Much of the work has defined how the cell interacts with the *cis*-acting viral regulatory sequences to induce the transcription of viral mRNA. In the case of HIV-1, the *cis*-acting sequences and *trans*-acting factors differ significantly from those in HIV-2. By studying these different processes in the virus, one gains a perspective for gene therapy. Viruses serve as a natural paradigm for what we do in gene therapy. Ordinarily, a virus would first infect a cell. Infection would be followed by a latent period, and latency would be succeeded by active expression of viral genes (inevitably toxic to the cell), allowing the propagation of the virus. When we seek to use genetic interventions for human disease, we attempt to deliver a gene that ordinarily would not be in the cell and then to regulate it so that the gene will produce therapeutic proteins within that cell. To deliver their genes efficiently, delivery must not only be targeted to the appropriate cell type but, once it has entered the cell, the virus must deliver the gene to the nucleus where it should be regulated in an appropriate fashion. Finally, the virus must be able to modulate immune function (in some cases the immune system must be avoided, and in other cases it may be necessary to stimulate it), as we must do in gene therapy.

2. Site-specific gene expression

When human disease is approached from the perspective of transcriptional regulation, it is clear that there is a parallel between basic questions about site-specific gene expression and those that are relevant to the treatment of human disease. Most diseases are important not only because there is an abnormality in gene expression, but also because of the specific sites where those abnormalities occur. Renal artery stenosis provides a graphic example. This disease produces focal abnormalities in blood vessels perfusing the kidney, which lead to systemic, often life-threatening, hypertension. Yet, only a small area of gene expression is affected. Consequently, a major goal of gene therapy is to target expression of therapeutic genes to specific sites. This entails the development of methods which target genes to specific sites *in vivo* as, for example, in the vessel wall. The arterial wall consists of three layers, an endothelial cell layer, smooth muscle and the adventitia, each separated by internal and external elastic lamina. Expression of a recombinant gene in the endothelium might be expected to have a very different effect from that which it would in the vascular smooth muscle layer. If a gene is expressed in the endothelium, the possibility exists of allowing it to be secreted and to perfuse a tissue that is distal to the site of introduction. In contrast, in a smooth muscle cell, it might be expected to exert a primary autocrine effect at the site at which the gene is expressed. The former has potential for drug delivery, while the latter might be used to modify the response to vessel injury.

A technique has been developed which uses catheter technology to deliver genes to specific sites *in vivo* (*Figure 1*). Two balloons are inflated within the blood vessel wall and these occlude blood flow to this region of the vessel. Between the balloons, a protected space is created. Vectors of various sorts, such as retroviral vectors, DNA–liposome complexes and adenoviruses, are used to target this specific area. Recombinant genes can be introduced into the vessel by exposing the genetic material to vascular cells. In cell-mediated gene transfer, a

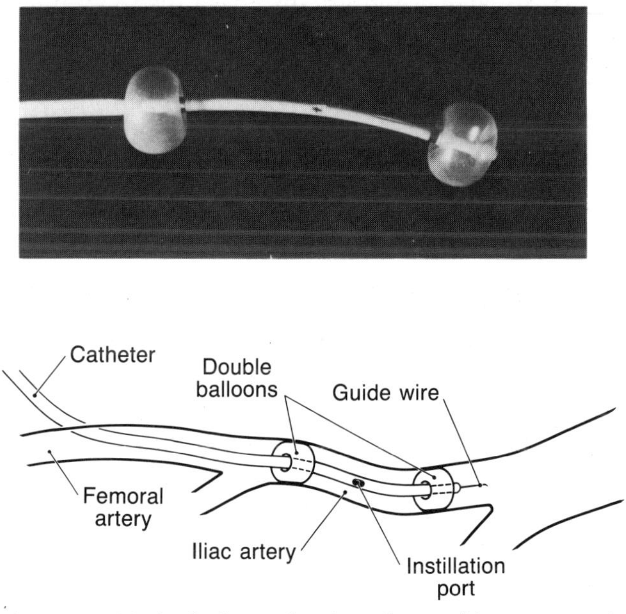

Figure 1. Method of introduction of recombinant gene. A double-balloon catheter is placed in the vessel. Inflation of the proximal and distal balloon creates a central protected space into which the modified cells or vector can be directly instilled. Vascular cells in the adjacent area are transfected to express the recombinant gene. Reprinted from Nabel *et al.* (1990) with permission from the American Association for the Advancement of Science. © 1990 by the AAAS.

segment of vessel is used to isolate endothelial cells or vascular smooth muscle cells, which are then transduced in the laboratory and reseeded through the instillation port of the catheter on to the artery *in vivo*.

Using vascular smooth muscle cells, non-transduced and transduced cells were reseeded on to arteries using a double-balloon catheter. Blue staining cells represent cells transfected with a β-galactosidase reporter gene (*Figure 2*). From a practical point of view, however, this technique of cell-mediated gene transfer is fairly cumbersome. Every time a new gene is introduced, cells must be grown in the laboratory, transduced and finally reseeded. Trying to apply this technology to human disease would be labour-intensive and, clinically, very difficult to apply. For that reason, direct gene transfer, instilling the retroviral vector or the DNA–liposome complexes into the vessel wall at specific sites, has been used. The advantage of this approach is its simplicity; the potential disadvantage is that some control is lost over the cell-specific expression because, once it is introduced, all cells take up the DNA. It is possible to deal with this problem by the design of cell-specific enhancers in the vectors, which allows introduction of the DNA into a variety of different cell types *in vivo*. The regulatory elements will allow expression only in specific cell types, for example in endothelial cells or vascular smooth muscle cells. Consequently, the ability to perform gene transfer directly and achieve cell-specific expression is possible.

Another possible technique is to occlude the arterial circulation using the distal balloon and then to transduce the distal capillary circulation. In fact, this technique can be used successfully in the distal femoral artery in the pig and the capillary endothelial cells can be transduced. The advantages of this method are that a large tissue bed can be perfused and the many endothelial cells in small capillaries present opportunities for systemic drug delivery through this cell type. At the same time, there is much that still needs to be understood about the basic biology of vessels *in vivo*: for example, what is involved in pathological processes, e.g. in restenosis, following catheter angioplasty for coronary artery disease.

3. Growth factor function *in vivo*

A research programme is under way to study the *in vivo* expression of specific recombinant genes in the arterial wall *in vivo*, for example platelet-derived growth factor B (PDGF B). From many *in vitro* studies, we know that PDGF B stimulates fibroblasts to divide. In the context of the native vessel, this technology provides the opportunity to create a

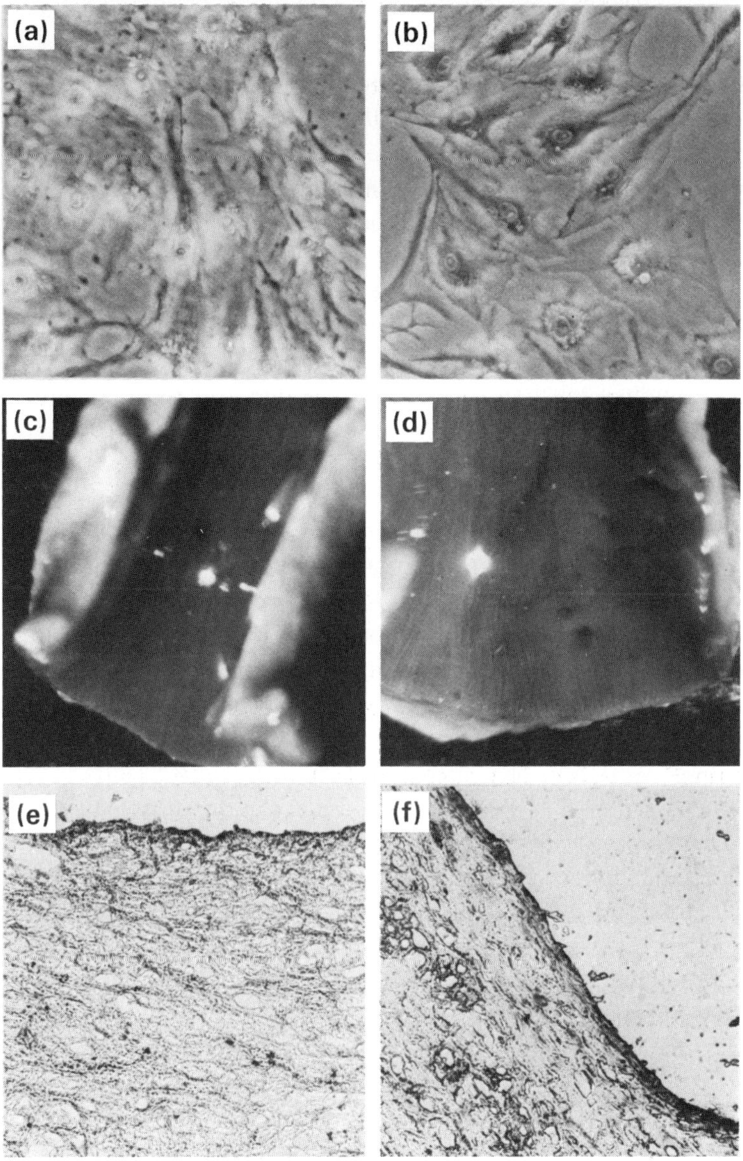

Figure 2. Analysis of gene transfer in endothelial cells *in vitro* and *in vivo*. β-Galactosidase activity was analysed by histochemical staining in: (a) primary endothelial cells from the Yucatan minipig, (b) a subline derived by infection with the BAG retroviral vector, (c) a segment of normal control artery, (d) a segment of artery instilled with endothelium infected with the BAG retroviral vector, (e) microscopic cross-section of normal control artery, and (f) microscopic cross-section of artery instilled with endothelium transduced with β-galactosidase. Reprinted from Nabel *et al.* (1989) with permission from the American Association for the Advancement of Science. © 1989 by the AAAS.

somatic transgenic animal for any number of genes to be studied and to evaluate the role and the activities of these genes *in vivo*. If the gene (DNA–liposome complex) is introduced, it is possible to verify its presence in transduced vessels by carrying out polymerase chain reaction (PCR) amplification of DNA from those arteries. It is not found in non-transduced arteries or in negative controls, but is detected in the positive control. Its presence can also be confirmed using reverse-transcription PCR, which shows that the signal from the RNA expressed in this segment of the vessel wall can be detected. This technique provides a mechanism to achieve expression of the gene in a site-specific fashion *in vivo*. It is also possible to document that the gene is being made at the DNA, RNA and protein level. This is now routine for almost all analyses. When using a vessel that has been transduced with a control gene, in this case β-galactosidase, a normal elastic lamina is observed. When PDGF B is transduced, there is proliferation of cells, but this could also be due to an effect on migration of cells and attraction of cells into the region where the gene is overexpressed (*Figure 3*). It is becoming clear from these kinds of studies that the mechanism of action may be more than a simple proliferative one.

An additional gene being investigated is *FGF-1*, formerly known as the gene encoding acidic fibroblast growth factor (FGF), which is an angiogenic factor. We have asked what role *FGF-1* plays in the pathology of the vessel if overexpressed *in vivo*. With *FGF-1*, proliferation of cells occurs within the intima, which suggests that not only PDGF but also *FGF-1* can lead to the expansion of this internal layer. This finding suggests that expansion of the internal layer might be achieved by different genes. With *FGF-1*, this proliferative phase of hyperplasia results. When the gene is substantially overexpressed, there is an early proliferative hyperplastic lesion, followed by the formation of capillaries. There are small vessels coursing through this expanded intima. A lumen running parallel to the lumen of the artery is seen, all within that expanded intima, something not seen with other non-angiogenic growth factors. Consequently, there is more polymorphism than expected within this intimal layer. Looking at the types of connective tissue which are made helps us to understand how these different growth factors might function in terms of vascular pathology. For example, one of the classic features of atherosclerosis is expanded intima as well as neovascularization of plaques that may help to keep that area of expansion perfused. It is likely that angiogenic factors such as *FGF-1* provide expansion *in vivo*, and at the same time have the ability to form new blood vessels *in vivo* in a site-specific way. There are obvious therapeutic implications that are now being developed. In one set of

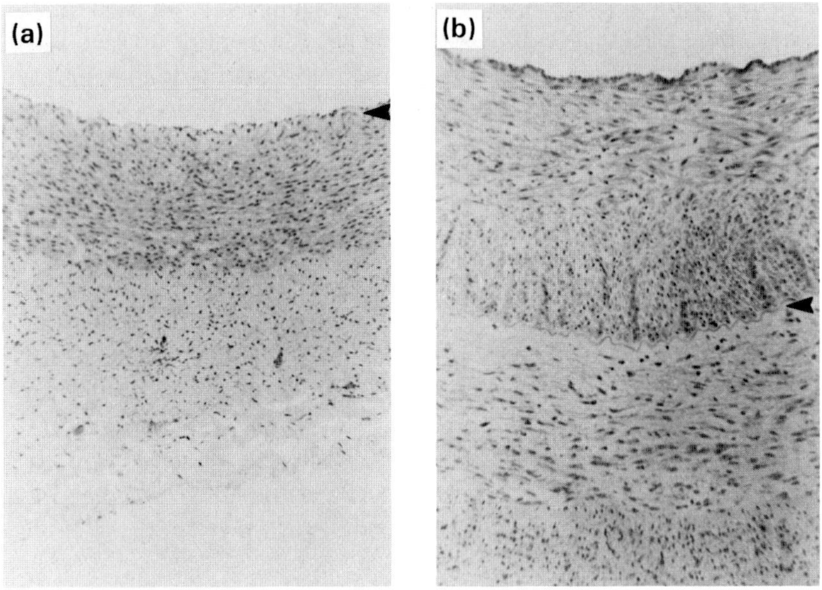

Figure 3. Expression of a recombinant growth factor gene in porcine arteries *in vivo*. Intimal thickening was produced following transfection of a human PDGF B gene into porcine iliofemoral arteries (b) compared with no intimal proliferation following transfection with a control gene, *E. coli*, β-galactosidase (a). The arrow denotes the internal elastic lamina. Magnification × 90. Reprinted from Nabel *et al.* (1993b) with permission from the American Society for Clinical Investigation.

experiments, endothelial cells were first transduced *in vitro*, and then reseeded into the subcutaneous tissue of nude mice. Cells were transduced with *FGF-1*, and overexpression induced the formation of new capillary beds. In fact, this strategy might provide a way of inducing collateral circulation in people who have end-stage coronary artery disease.

4. Immunotherapy by direct gene transfer *in vivo*

After discovering that β-galactosidase could be expressed in a vessel wall, it was important to demonstrate not only that vessels can be stained with a chromagen, but also that the biology of the vessel could be changed *in vivo*. One of the most sensitive biological reactions is the transplantation response to a foreign class 1 histocompatibility gene, which is largely responsible for transplant rejection. A foreign major histocompatibility complex (MHC) gene is composed of an α-1 and α-2 helical domain in the region that presents the peptide to the T-cell receptor linked to the α-3 domain, paired to β_2-microglobulin through disulphide bonds. In initial experiments, retroviral vector or DNA–liposome complexes encoding this gene were introduced.

The same techniques were used to confirm expression at the DNA, RNA and protein levels (*Figure 4*). Inflammatory cells are found at the site where the gene was expressed. Control-transduced vessels remained essentially normal, but expansion was observed in the adventitia of this arterial segment, and there was considerable expansion of the intimal layer. Also, the media were more engorged and disrupted. Essentially, this genetic modification induced a granuloma-

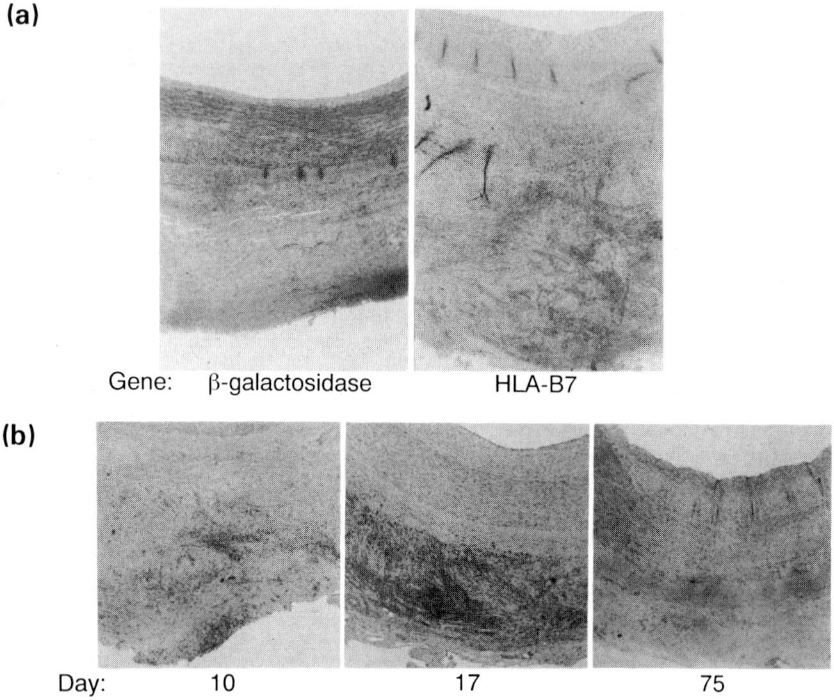

Figure 4. Pathology of the inflammatory reaction produced by the expression of recombinant human *HLA-B7* gene in porcine vascular cells *in vivo*. (a) Porcine arteries transduced with a reporter gene, *E. coli* β-galactosidase (left) or human *HLA-B7* gene (right) 17 days after transfection. (b) Time course of inflammatory response in the adventitia, media and intima 10 (left), 17 (middle) and 75 (right) days following transfection. Magnification × 80. Reprinted from Nabel *et al.* (1992) with permission from the National Academy of Sciences.

tous response, consisting of T cells and macrophages. This type of pathological picture is similar to that encountered in Takayasu's vasculitis, an autoimmune disease of unknown aetiology that affects young women. An immune response that exposes antigens in the vessel wall (normally not seen by the immune system) seems to have been induced and, by exposing these antigens to the immune systems, a secondary proliferative response occurs. A high-powered view of that vessel reveals its pathology, typical of the changes seen in a granulomatous vasculitis: the infiltration of mononuclear cells is visible, and an 'onion skinning' effect is present around the small capillaries. In a controlled transduced vessel, a normal single cell layer is seen around the small capillaries. In *HLA-B7* vessels, these small capillaries are obliterated. Another non-transduced artery from the same animal shows no pathology. The inflammation is localized specifically to the site where the gene has been introduced. These animals generate cytolytic T cells which are capable of lysing cells from the pig that expresses *HLA-B7*, but not those that do not. Consequently, a specific cytolytic T-cell response can be induced that is specific for the recombinant gene that was introduced. It is also localized specifically to the site where it was introduced. This experiment also shows that a cytolytic T-cell response can be generated by producing the product of the gene that was introduced *in vivo*. In fact, this seems to be a particularly good way of inducing cytolytic T cells. It is possible to generate cytolytic T cells against the envelope protein for HIV. In a similar fashion, producing a cytolytic T-cell response against gp160 or gp120 by introducing the gene stimulates the immune system in a different way. In terms of therapeutic vaccines for HIV, this could be of great importance.

To develop this model further, transplantable tumours were used which necessitated changing to the murine model. Instead of using *HLA-B7*, a class 1 gene that was allogenic to the host strain, the *H-2K* gene, was used. This follows the same principle as was previously described, with a different gene, involving introduction of a foreign haplotype to the mouse. The model is one where the tumour is inoculated subcutaneously, and then grows. Direct intratumour injection is used to deliver the MHC gene, induce an immune response and to determine whether it has a protective effect for cancer. The introduction of this gene into the CT26 colon carcinoma was accomplished with either retroviral vectors or DNA–liposome complexes. β-Galactosidase was examined as a control gene. The ability to generate T cells that could lyse chromium-labelled CT26 (the parental tumour line) or those modified with the same gene was examined. Injection of the relevant gene induced an immune response against either cells modified by the s haplotype of *H-2K* (K^s) or against the parental cell. Expression of the foreign gene in the tumour target allowed generation of a cytolytic T-cell response against the product of the gene implanted. What does this do to the immune system's ability to recognize the parental unmodified tumour cells? In contrast to the control gene, it is now possible to generate a cytolytic T-cell response against tumour-associated antigens. The response to both K^s and CT26 is T-cell-mediated and blocked by anti-lyt2, or the murine equivalent of CD8. This effect can protect against tumour growth in recipient animals. In 70% of animals, it results in a 50% reduction in tumour diameter 3 weeks after the treatment. Tumours treated with β-galactosidase have shown no response to treatment. This approach is potentially universal. If this gene is injected into the MCA 106 fibrosarcoma, the immune system is also induced to recognize this tumour, but not a third-party tumour. If the same is done in CT26, a colon carcinoma, the immune system is induced to recognize it but not MCA 106. The immune system has been induced to respond to its antigens. This approach can, therefore, be applied to a variety of different tumour cells.

Although this technique is not uniformly curative at this time, available data suggest that the life span of animals is doubled (achieving permanent cures in a little more than 20% of animals). The procedure must be refined; however, murine tumours are highly inbred, extremely aggressive in their growth, and to provide even this amount of protection is remarkable. In some instances, we have found animals that remain tumour free for more than 150 days. The control animals died within approximately 1 month. Of course, questions remain: is this procedure anything more than molecular surgery? Will the technique help to treat cancer *in vivo*, and be effective at distant sites? To answer such questions, a micrometastatic model was developed using the B16 melanoma cell line. Animal tumours were treated with an intratumour injection, with β-galactosidase or K^s. The draining lymph nodes from injected animals were used to inoculate recipient animals 3 days after intravenous injection with B16. A highly significant reduction in the number of lung nodules was seen in K^s-treated animals. When 10^7 lymphocytes were administered per animal, nearly complete protection was observed, illustrating the power of the immune system. Based on these animal models, as well as a number of toxicity studies and product development, this protocol was adapted for the treatment of human disease. In the initial phases, the mode of delivery (intratumour injection) was preserved, although eventually the genes could be delivered into the tumours by catheter.

The potential advantages of this method of gene transfer are: (i) simplicity; (ii) low toxicity (in the animal models); and, most importantly, (iii) sensitive and accessible indicators of recombinant gene expression *in vivo*, since the expression of the gene is self-limited. For example, putting cytokines into tumours without being able to regulate them could compromise safety, because many of them can be growth stimulatory and unregulated expression of a growth stimulatory gene is best avoided. Patients with stage 4 melanoma, who have no treatment options, generally fail even immunotherapy before undergoing the treatment described herein. There are potential disadvantages, of course. For one, the long-term effects of expression of these genes *in vivo* is unknown and this is one reason for conducting a human clinical trial. The vector introduced is remarkably simple: a Rous sarcoma virus enhancer regulates expression of *HLA B7*, using the SV40 splice and poly(A) sequences on a pBR322 backbone.

The trial design is typical for a phase 1 protocol, which involves injecting increased amounts of DNA. The first patients received a single injection of the DNA–liposome complex into the nodule, repeated at 2-week intervals for a total of three injections, and with the total amount of 0.86 μg of DNA. The next group of patients received three injections per nodule three times, followed by five injections per nodule for three treatments and, lastly, five injections for five treatments. The first half of this trial has been completed but it will probably not be continued as most of the relevant information has been obtained. What is there yet to learn? Firstly, is the gene being expressed? Secondly, is there any evidence of toxicity? Thirdly, is there an immune response to *HLA-B7*? Finally, is there an immune response to the tumour?

In terms of gene therapy of malignancy, the obvious difference between the described approach (i.e. direct gene transfer) and those described previously is that the direct gene transfer method is simpler; furthermore, treatment can begin soon after diagnosis. In the case of cell-mediated transfer, there is much that needs to be done before patients can be treated: the diagnosis; the extraction of the tumour; the ability to grow the tumour in the laboratory; the ability to transduce the tumour; and, finally, readministration to the patient.

5. Conclusions

In summary, site-specific expression of recombinant genes can be achieved by direct gene transfer *in vivo*, which simplifies the application of molecular biology to the treatment of human disease. In the specific case of the MHC gene, the immune system can be induced to recognize antigens which it previously did not recognize. The clinical protocol has been initiated and the process begun of gathering information that will be relevant to future investigations into a variety of human diseases. Furthermore, if the safety of catheter-based gene delivery in a cancer model can be demonstrated, eventually it may become possible to apply the method to the treatment of cardiovascular diseases.

It is now possible to think about utilizing molecular genetic interventions for human disease in ways that have not been used before. For those who try to understand the basic mechanisms of pathogenesis and biology, this represents an opportunity to use knowledge derived from molecular genetics in a rational way and, in the case of malignancy, to begin to target multiple gene products that may be causally involved in tumour pathogenesis. If approaches can be further developed that enable genes to be delivered at specific sites, thus affecting multiple steps in the processes that lead to oncogenesis (growth factor receptor genes, oncogenic proteins, and dominant recessive oncogenic proteins) multiple factors that regulate tumour growth can also be targeted. For human diseases, this methodology will allow a better understanding of the diseases at a molecular level, and also allow knowledge to be translated in a meaningful, rational way, providing new tools to develop treatments for incurable diseases.

References

Nabel EG, Plautz G, Boyce FM, Stanley JC and Nabel GJ. (1989) Recombinant gene expression *in vivo* within endothelial cells of the arterial wall. *Science*, **244**, 1342–1344.

Nabel EG, Plautz G and Nabel GJ. (1990) Site-specific gene expression *in vivo* by direct gene transfer into the arterial wall. *Science*, **249**, 1285–1288.

Nabel EG, Plautz G and Nabel GJ. (1992) Transduction of a foreign histocompatibility gene into the arterial wall induces vasculitis. *Proc. Natl Acad. Sci. USA*, **89**, 5157–5161.

Nabel GJ, Nabel EG, Yang Z, Fox B, Plautz G, Gao X, Huang L, Shu S, Gordon D and Chang AE. (1993a) Direct gene transfer with DNA liposome complexes in melanoma: expression, biologic activity and lack of toxicity in humans. *Proc. Natl Acad. Sci. USA*, **90**, 11307–11311.

Nabel EG, Yang Z, Liptay S, San H, Gordon G, Haudenschild CC and Nabel GJ. (1993b) Recombinant platelet-derived growth factor B gene expression induces intimal hyperplasia *in vivo*. *J. Clin. Invest.* **91**, 1822–1829.

Plautz GE, Yang ZY, Wu B, Gao X, Huang L and Nabel GJ. (1993) Immunotherapy of malignancy by *in vivo* gene transfer into tumors. *Proc. Natl Acad. Sci. USA*, **90**, 4645–4649.

12

Processing of membrane proteins in neurodegenerative diseases

R.J. Mayer[1], C. Tipler[1], M. Landon[1] and J. Lowe[2]
Departments of [1]Biochemistry and [2]Pathology, University of Nottingham Medical School, Nottingham, UK

1. Introduction

Demographic predictions indicate increased numbers of old people in the 21st century; the social and economic implications of neurodegenerative disorders associated with ageing will hence become ever more significant. The problems of age-related infirmity are succinctly illustrated by an obvious increase in the number of homes for the elderly throughout Westernized countries, although changing patterns of family behaviour have also contributed to this growth.

The human neurodegenerative diseases can be categorized into idiopathic and transmissible disorders: the former have a far higher incidence and include Alzheimer's disease (AD), Lewy body dementia (LBD) and Parkinson's disease (PD); the latter include Creutzfeldt–Jakob disease (CJD), Gerstmann–Straussler–Scheinker disease (GSS) and fatal familial insomnia.

AD and related disorders, such as LBD, exhibit the commonality of amyloid plaques. Amyloids are extracellular fibrillar materials in which proteins or protein fragments are associated by protein secondary structural interactions based on β-sheet formation. In AD-related disorders, the extracellular amyloid contains, as a major component, aggregates of a fragment, Aβ (previously referred to as β-protein or β/A4), of the Alzheimer amyloid precursor protein (APP). The functions of this membrane protein remain unknown. Other pathological features include either neurofibrillary tangles in AD or cortical Lewy bodies in LBD.

In addition to the human transmissible diseases, there are related animal disorders including sheep scrapie and bovine spongiform encephalopathy (BSE). The weight of scientific evidence indicates that, contrary to the central dogma of molecular biology, replication of the causative agents of transmissible diseases, prions (from proteinaceous infectious agent; Prusiner, 1991), occurs in the absence of a nucleic acid (Prusiner, 1991, 1992). The prion diseases, a description that is increasingly used as an alternative to spongiform encephalopathies, may be both transmissible and genetic: site(s) for, and mechanisms of replication of, prions are the subjects of much current research into the disorders. The outbreak of BSE in the UK has highlighted the transmissible diseases, both scientifically and in the media. Iatrogenic transmission of prions in pituitary-based preparations of growth hormone and gonadotrophins to thousands of patients is a clinical and political time-bomb, while the possibility of transmission by blood transfusion complicates and extends the spectre.

Prion diseases may also exhibit extracellular amyloid, where the aggregates are composed not of fragments of APP, but of the prion protein (PrP), otherwise called the protease-resistant protein because of the way the abnormal form of the protein in diseased brain is identified through its resistance to digestion with Proteinase K. PrP is a glypiated protein (phosphatidyl inositol glycan is covalently attached to the protein to provide a membrane anchor), which accumulates predominately in a non-truncated form in the aggregates of extracellular amyloid (Prusiner, 1991, 1992). As with APP, the functions of PrP remain to be established.

The genetics of AD-related disorders and the prion diseases point to age-related problems in the post-translational processing of APP and PrP, following insertion into cell membranes, as central to disease pathogenesis (Mayer *et al.*, 1991b, 1992; Prusiner, 1991, 1992). Key unknowns are the site(s) and molecular mechanisms by which the proteins are converted into their amyloid forms.

Unexpectedly, the accumulating molecular cell biological evidence supports the notion that the processing events in the generation of amyloidogenic versions of the proteins occur after endocytosis of the proteins in endosome–lysosomes (Borchelt *et al.*, 1992; Laszlo *et al.*,

1992; Mayer *et al.*, 1992; McKinley *et al.*, 1991). Presumably, APP and PrP enter this pathway as part of the normal process leading to their degradation to amino acids. In disease pathogenesis, fragmentation of the proteins is incomplete or blocked, which results in the accumulation of aggregates of fragments of APP or intact PrP, respectively. Subsequently, amyloid is expelled from the cell by fusion of endosome–lysosomes at the cell surface, following the accumulation of increasing quantities of extracellular materials. Regurgitation of surface proteins from the endosome–lysosome system is a natural cellular process, e.g. in the expulsion of transferrin receptors from maturing reticulocytes (Davis *et al.*, 1986), and can therefore be anticipated to have a key role in the ejection of amyloid materials from susceptible cells in the neurodegenerative diseases.

The concept of endosome–lysosomes serving as 'bioreactors' for the generation of amyloidogenic protein aggregates in neurodegenerative diseases serves as a central unifying hypothesis to explain the diseases as abnormalities in the processing of single membrane proteins (Laszlo *et al.*, 1992).

2. Idiopathic neurodegenerative diseases

The disease which carries his name was first described by Alois Alzheimer early this century and is associated with declining cognitive capacity of varying duration. For many years use of the term AD was restricted to pre-senile dementia, that is, in persons under 65 years of age. The much more common occurrence of the disease in persons over 65, and awareness that the neuropathology of early and late cases was similar, suggested that the underlying molecular causes of pre-senile and senile dementia might be the same. Thus, the age distinction was dropped and the definition of AD encompassed both early- and late-onset forms. Current knowledge suggests that AD is more heterogeneous than was suspected.

Another conceptual advance is that many patients presenting with dementing illness have LBD rather than AD (Byrne *et al.*, 1991; Hansen *et al.*, 1989; Lennox *et al.*, 1989a). LBD was originally named diffuse Lewy body disease (Okazaki *et al.*, 1961) and was seen as very rare: now recognized as more common, it is also called senile dementia of the Lewy body type (Perry *et al.*, 1989) or the Lewy variant of AD (Hansen *et al.*, 1990). LBD is distinguishable clinically by rapid stepwise patient deterioration, often accompanied by hallucinations and behavioural disturbances. Neuropathologically, AD and LBD both exhibit extracellular plaques but they differ in the nature of the intraneuronal inclusions that are seen: intraneuronal neurofibrillary tangles in AD and cortical Lewy bodies in LBD. The neurofibrillary tangles are composed predominantly of hyperphosphorylated tau (Brion *et al.*, 1991), a microtubule-associated protein (MAP). Cortical Lewy bodies contain predominantly hyperphosphorylated neurofilament proteins (Schmidt *et al.*, 1991). In rare cases, both intraneuronal neurofibrillary tangles and Lewy bodies occur, indicating mixed double pathology (Lowe *et al.*, 1993).

Cortical Lewy bodies have become routinely detectable only with the advent of ubiquitin immunocytochemistry (Byrne *et al.*, 1991; Lennox *et al.*, 1989b). Clearly, cortical Lewy bodies which, like neurofibrillary tangles, are immunologically ubiquitin-positive, must contain some form(s) of ubiquitinated proteins, the molecular nature of which is currently unknown. Brain stem Lewy bodies which are pathognomonic for PD have been described for many years. Lewy bodies are found widely distributed in cortical areas in LBD and their numbers correlate with the clinically determined degree of dementia (Lennox *et al.*, 1989b). It has been speculated, based on the distribution of some molecular markers in neurofibrillary tangles and Lewy bodies (PGP 9.5, a ubiquitin carboxy-terminal hydrolase; αB crystallin and gelsolin), that tangles represent 'tombstone' inclusions in dying or dead cells. In contrast, Lewy bodies may form in neurones as part of an active process in which neurofilaments are used in an attempt to 'cocoon' unwanted or lethal proteins or organelles prior to protein degradation by the ATP and ubiquitin-dependent non-lysosomal system or in the lysosomal apparatus (Lowe and Mayer, 1990; Lowe *et al.*, 1993; Mayer *et al.*, 1991a,b).

2.1. The APP gene and protein products

The gene for APP resides on chromosome 21 and consists of 19 exons. It codes for several glycoproteins, with the different isoforms being generated by alternative splicing of exons 7 and 8 (which code for, respectively, the Kunitz protease inhibitor domain and the MRC OX2 thymocyte antigen homologous domain). The different forms of APP produced include APP695, APP714, APP751, APP770 and APRP563. A further APP variant, leukocyte-derived APP (L-APP) has also been discovered; this lacks exon 15 of the APP gene and is expressed by T lymphocytes, macrophages and microglial cells (Konig *et al.*, 1992). Relationships between the gene and gene products are shown in *Figure 1*.

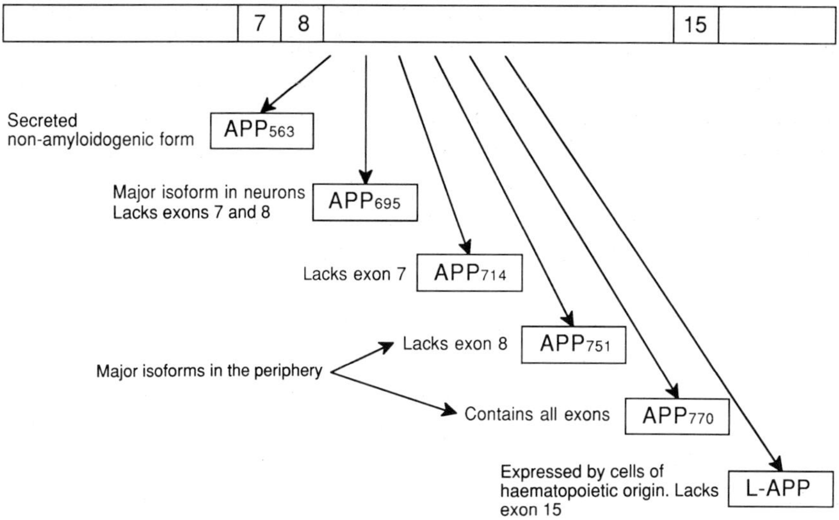

Figure 1. Alternative transcripts of the Alzheimer's precursor protein gene. There are 19 exons – only those involved in alternative splicing are indicated.

APP695 mRNA is the major type found in neurons, while the larger forms (which contain exon 7) are predominantly expressed in peripheral tissues. The amyloid fragments, Aβ, are the 39–43 amino acids derived from part of the single membrane-spanning domain of APP plus amino-terminal residues from the region of the protein abutting on to the membrane (see *Figure 2*).

The APP protein can be physiologically cleaved within the Aβ sequence to release the large amino-terminal fragment of APP (Esch *et al.*, 1990), an event that prevents the formation of Aβ and ensuing amyloid deposition. This cleavage in Aβ (at the carboxy-terminal side of residue 16; see *Figure 2*) is catalysed by an incompletely characterized α-secretase enzyme(s), possibly within the secretory pathway or at the cell surface. Another enzyme(s), called β-secretase, can cleave precisely at the amino-terminus of Aβ (Seubert *et al.*, 1993). The released amino-terminal fragments of APP are presumed to have some physiological functions outside the cell (Oltersdorf *et al.*, 1989). These could include inhibition of proteases, or some regulatory function (e.g. in the control of neuronal growth, differentiation and nerve process extension, for example, in response to neurodegeneration).

The enzymes cleaving at the carboxy-terminus of Aβ are not fastidious, releasing several fragments with various carboxy-termini.

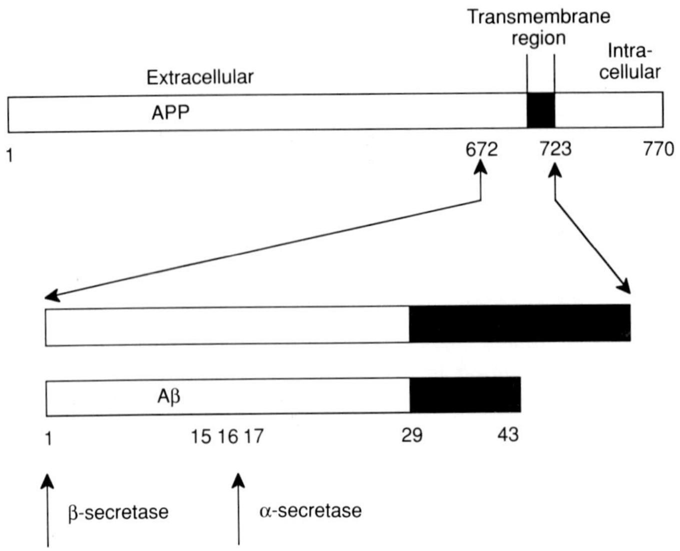

Figure 2. Generation of the Aβ peptide from the Alzheimer's precursor protein.

Currently there is considerable interest in the pathological significance of these different forms. Thus, while Aβ1–40 is normally produced in the brain in small amounts and appears in the cerebrospinal fluid (Selkoe, 1991), its relevance to pathological deposits is unclear. This uncertainty is reinforced by recent proposals that the predominant species in vessels (Roher *et al.*, 1993) and plaque cores is Aβ1–42 (Roher *et al.*, 1993). Furthermore, *in vitro* studies show that Aβ molecules containing up to 40 residues are apparently much less likely to form insoluble material than those that are two or three residues longer, and also that Aβ1–43 accelerates the precipitation of Aβ1–39 (Jarrett *et al.*, 1993). Whatever the form, or forms, of Aβ required for disease-related accumulation of insoluble molecules in the brain, it seems likely that there will be age-related differences in their rates of production.

2.2. Mutations in the APP gene

Identification in the coding region of the APP gene (*Figure 3*) of several different mutations that are linked with early-onset AD, has served to focus on the importance of APP and amyloid deposition in disease pathogenesis (Price *et al.*, 1993). Although a familial connection with AD has been known for many years, finding APP mutations in certain families has made the APP protein central to the occurrence of disease.

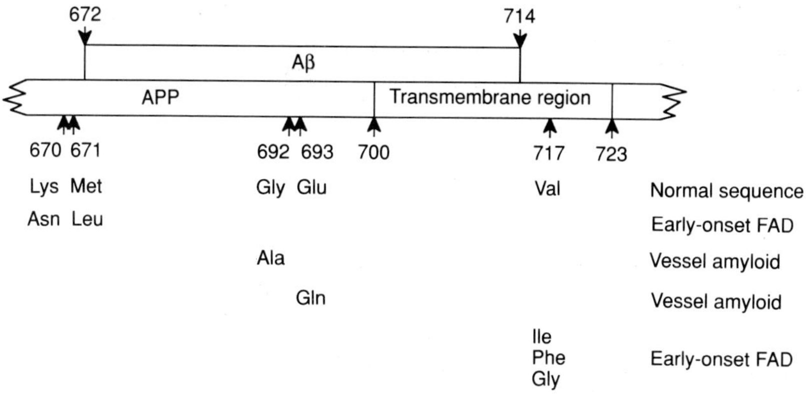

Figure 3. Mutations in the Alzheimer's precursor protein gene associated with disease.

Several of the mutations in the coding region of the APP gene are within (normally associated with amyloid deposits in vessels rather than AD *per se*) or around the Aβ sequence and presumably influence the catabolism of the mutant APPs, resulting in the enhanced accumulation of Aβ. Persons with mutations in the APP gene account for only a small number of those with early-onset familial AD (FAD), a majority of whom have linkage with a currently unidentified gene on chromosome 14.

2.3. Functions of APP

The precise functions of APP are unknown. While the predicted amino acid sequence suggested that APP could be a transmembranous receptor, there is currently no evidence for this notion. Alternatively, APP may be the substrate for the α-secretase in the secretory pathway or at the cell surface, which releases the large amino terminal fragment of APP as a substance functionally active in the regulation of cellular homeostasis in the nervous system and elsewhere in the organism, for example in the platelet release reaction. There is some evidence for the binding of the secreted form of APP to extracellular glycosaminoglycans, from where the fragment could be released slowly for some physiological purpose, in a manner similar to that for fibroblast growth factor. The amount of APP is found to be increased after certain head injuries (Roberts *et al.*, 1994). Given that the occurrence of some cases of AD may be the consequence of brain trauma (e.g. in relation to dementia pugilistica or serious head injuries), APP may function to restore nerve function in damaged areas of the brain.

2.4. Catabolism of APP and generation of Aβ

Experimental evidence indicates that all membrane proteins are internalized into the endosome–lysosome system for degradation to amino acids (Mayer *et al.*, 1991). Endosome–lysosomes represent a family of related organelles in which membrane proteins, including receptor–ligand complexes, can be internalized. Endocytosis of cell surface membrane via clathrin-coated pits is followed by release of the clathrin-related components to give intracellular vesicles and tubules which can be involved in fusion events and maturation processes through early and late endosomes to give finally late, classical 'de Duvian', dense lysosomes. The maturation process is accompanied by a gradual acidification of the interior of organelles from neutrality, through pH 6.5–6.0 in early structures, to around pH 3.5–4.5 in the late lysosomes (Parton and Dotti, 1993). Acidification can drive receptor–ligand dissociation, allowing receptor recycling to the cell surface and protein–ligand degradation to amino acids. Although still incompletely characterized and probably variable between cell types, the internal milieu of the maturing organelles may be either reducing or oxidizing (see, for example, Doherty *et al.*, 1989a,b). Irrespective of the nature of the redox state, proteins entering these organelles will become unfolded and therefore rendered more susceptible to the battery of hydrolases which are loaded into the maturing organelles from hydrolase-transport vesicles that bud off from the trans-Golgi network and fuse with the maturing endosome–lysosomes. The hydrolases work best at acid pH and include proteases, phospholipases, glycosidases and other hydrolytic enzymes.

APP occurs in the endosome–lysosome system of cultured neurones (Benowitz *et al.*, 1989) and presumably, therefore, in neurones *in vivo*. There is also mounting evidence that a variety of Aβ-related peptides (perhaps up to a dozen variants) can be found in these organelles (Haass *et al.*, 1992; Siman *et al.*, 1993). The cellular site(s) of an enzyme(s) cutting at the carboxy-terminus of Aβ is unknown, but appears likely to be in the endosome–lysosome system. Recent studies indicate that cathepsin D-type enzymes may be involved in the generation of Aβ-related peptides, while the thiol-cathepsins (e.g. cathepsins B, H and L) may be involved in the further degradation to amino acids of the frayed Aβ-related molecules (Siman *et al.*, 1993). The proteolytic steps in the degradation of a specific protein in the endosome–lysosome system have never been fully characterized, probably because of the possibility of alternative pathways of proteolysis by different proteases in the endosome–lysosome system and, additionally, because the exponential-like process of proteolysis restricts the accumulation of measurable intermediates. Age-related increases in the surface concentration of APP or changes in the fluidity of the nerve cell membranes, the rate of internalization and recycling to the cell surface of APP, the pH of the organelles, the propensity for protein denaturation and the proteolytic and other hydrolytic capacities of the organelles could alter the steady-state concentration of the Aβ-related peptides and affect their accumulation. Such changes might favour larger and more amyloidogenic species (e.g. Aβ1–43) or the nucleation of the formation of Aβ-containing amyloid precursors which, after regurgitation at the cell surface, would accumulate extracellularly in the variety of forms of plaque. The proposed central role of endosome–lysosomes as 'bioreactors' for the formation of amyloid in AD is shown in *Figure 4*.

Any hypothesis purporting to explain amyloid generation in AD and related disorders must also be able to explain, either directly or indirectly, the concurrent intraneuronal cytoskeletal changes, namely neurofibrillary tangles or cortical Lewy bodies (*Figure 5*).

There is no convenient cellular or animal model for AD, although transgenesis may hold the answer to the search for suitable animal models. Furthermore, in so far as comparisons with other diseases can be made, there is no complete counterpart for the cellular changes seen

---

Figure 4 (see pp. 214 and 215). Proposed role of endosome–lysosomes in the carboxy-terminal processing of the Alzheimer's precursor protein. N, N terminus; C, C terminus. (a) Cleavage of APP by either α- or β-secretase to remove the amino-terminal fragment of the protein; (b) invagination of the cell membrane occurs. The proteins are endocytosed into the endosome–lysosome system; (c) the vesicles containing APP fuse with others to form multivesicular endosomes. Other vesicles have budded off from the trans-Golgi to ferry hydrolytic enzymes to the multivesicular bodies, including acid phospholipases which disrupt the phospholipid membrane of the internal vesicles; (d) the vesicles begin to mature, there is a further drop in pH, and acid proteases begin to digest proteins. The α-secretase product is fully degraded to small peptides and amino acids, but only limited proteolysis of the β-secretase product can occur, resulting in Aβ-related peptides with frayed carboxy termini which may be prone to aggregation in the denaturing environment of the organelles; (e) The Aβ-fragment aggregates contained in the endosome system are recycled back to the cell surface (cf. transferrin receptor). On fusion of the vesicle with the cell surface, the aggregates are released to produce extracellular amyloid deposits.

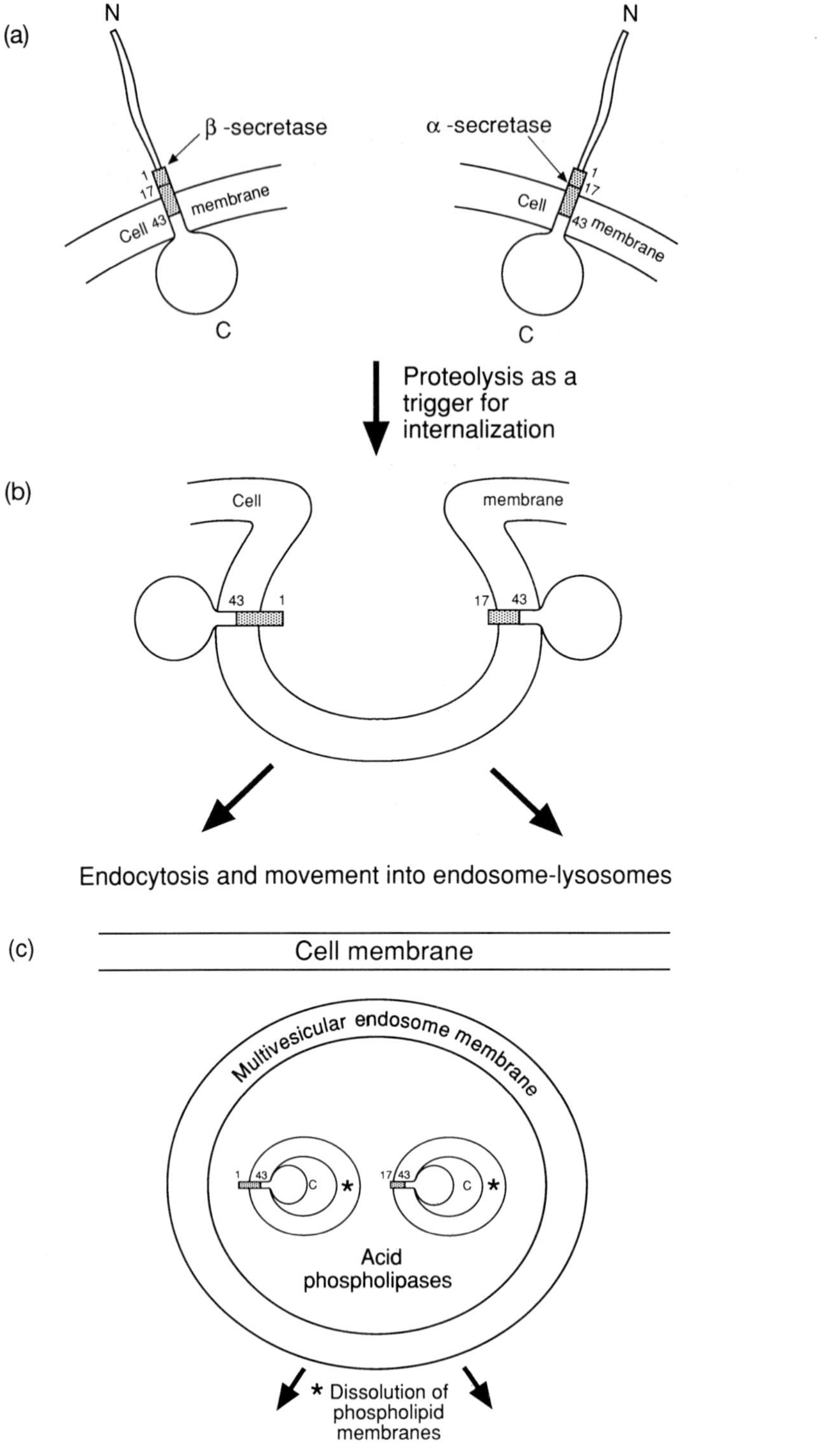
(a)
N
β -secretase
1
17
43
Cell membrane
C
N
α -secretase
1
17
43
Cell membrane
C
Proteolysis as a trigger for internalization
(b)
Cell membrane
43
1
17
43
Endocytosis and movement into endosome-lysosomes
(c)
Cell membrane
Multivesicular endosome membrane
1
43
C
*
17
43
C
*
Acid phospholipases
* Dissolution of phospholipid membranes

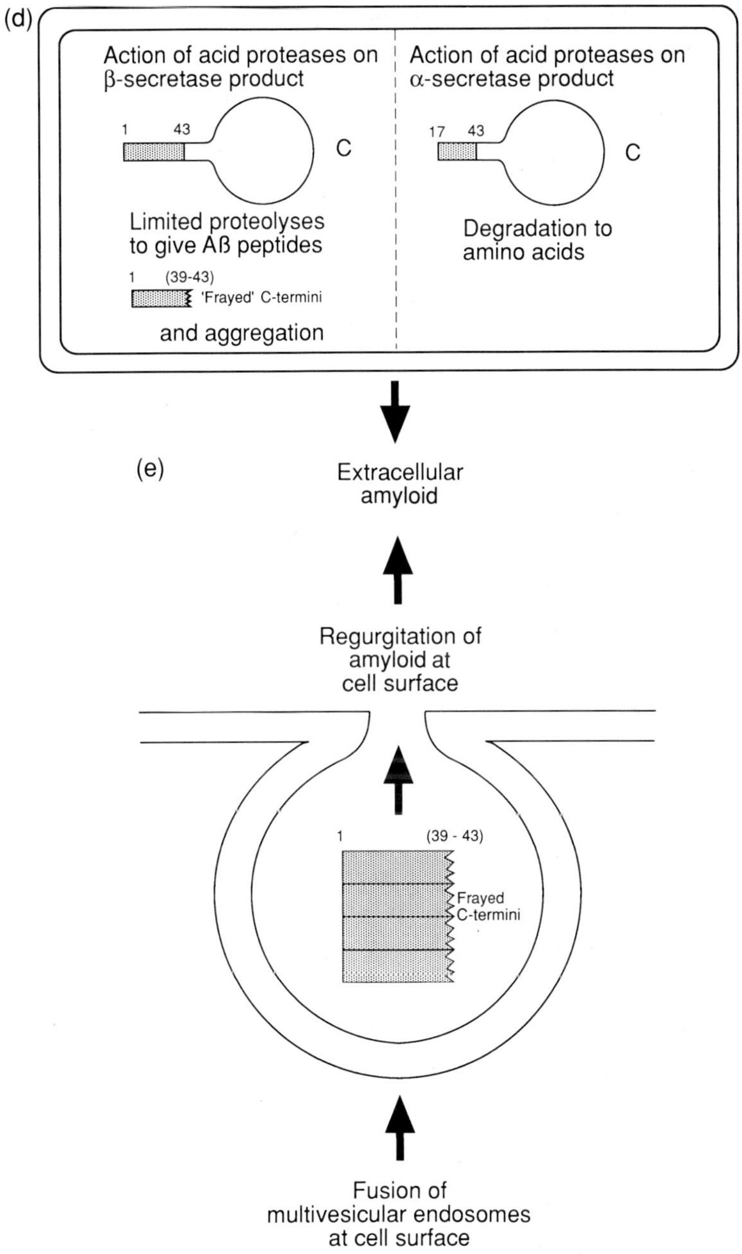

Figure 4 (cont'd).

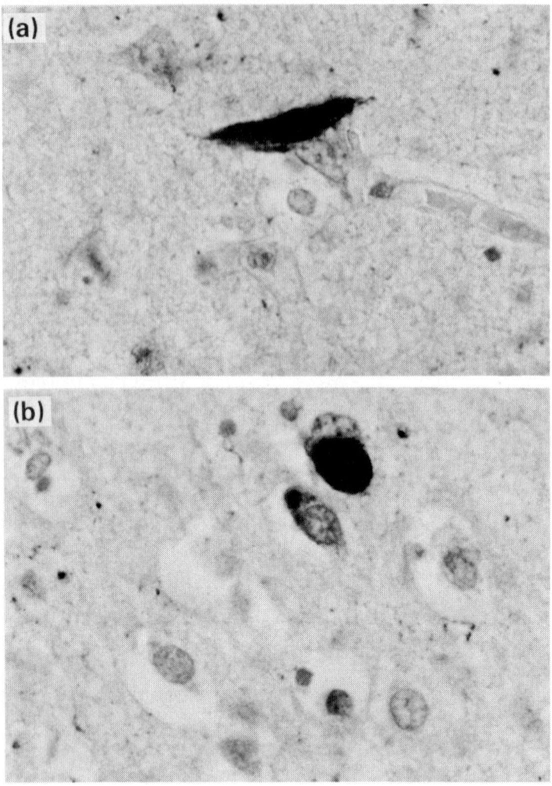

Figure 5. Neurofibrillary tangles in AD (a) and Lewy bodies in LBD (b) decorated with antibodies to ubiquitin–protein conjugates.

in the neurodegenerative diseases. However, studies on Epstein–Barr virus-transformed lymphoblastoid cells seem to provide some clues. Immunocytochemistry studies show that lymphocyte transformation by the virus or transfection of cells with the gene for the latent membrane protein (LMP, a multiple membrane-spanning protein of the virus that is responsible for tumorigenesis in susceptible animal cells) cause a reorganization of the lymphocyte cytoskeleton to produce a vimentin-based filament inclusion which also contains LMP, a heat-shock protein 70 (hsp 70) and ubiquitin–protein conjugates (vimentin is the intermediate filament protein in cells of mesenchymal origin). The inclusions in part resemble Lewy bodies, which contain neurofilaments (the intermediate filament protein in neurons) and ubiquitin–protein conjugates, together with several other proteins.

Ultrastructural investigation by immunogold electron microscopy shows that viral transformation of lymphocytes causes reorganization of the endosome–lysosome system to orientate large lysosome-related organelles, containing LMP, hsp 70 and ubiquitin–protein conjugates, around the microtubule-organizing centre of the cells. Chronic adaptation of transformed (and transfected cells) to the production of a viral membrane protein thus results in the collapse of the intermediate filament cytoskeleton to produce a filamentous inclusion. This is associated with large endosome–lysosome-related organelles, within which LMP is presumably being destroyed or ejected from the cell (Laszlo *et al.*, 1991b). Rapid reversible intermediate filament collapse can be triggered by the experimental loading of Sendai viral membrane proteins into cells by fusion at the cell surface before the degradation of the viral proteins in the endosomal–lysosomal system (Earl *et al.*, 1987). The cytomorphological changes seen in response to viral membrane proteins resemble, but are not exactly the same as, those seen in LBD. Clearly, cytoskeletal reorganization can occur acutely in response to loading of membrane proteins into the cell surface, and chronically in cells persistently infected with a virus that directs production of its own membrane protein. Similarly, problems in the handling of a single membrane protein in AD-related disorders may trigger cytoskeletal and endosome–lysosome rearrangements in neurones and other cell types.

Although the endosome–lysosome system of neurones is poorly characterized, in general, these organelles are found predominantly in nerve processes (with more in the nerve terminals than in the dendrites), while the 'de Duvian' matured lysosomes are confined to nerve cell bodies (Parton *et al.*, 1992). Endosome-like organelles experimentally transfer endocytosed molecules in a retrograde manner back from processes towards the cell body, presumably for both signalling to the nucleus (e.g. nerve growth factor) and for final degradation of protein and other ligands (Parton *et al.*, 1992). It has proved difficult to detect many endosome-like organelles in nerve processes in normal adult brain *in vivo*: in contrast, late lysosomes are readily seen in neuronal cell bodies by, for example, immunohistochemical detection of lysosomal cathepsins.

In AD-related disorders endosome–lysosome-like organelles can be detected at three sites by ubiquitin immunocytochemistry, using the light microscope: in dystrophic neurites surrounding amyloid plaques, in axosomatic processes abutting on to neurones (or in autophagic granules within certain neurones, e.g. those of the hippocampus), and as dot like structures in the neuropil (Lowe *et al.*, 1993). The dot-like structures in the neuropil are also a feature of the normal ageing brain in the absence of AD pathology (Dickson *et al.*, 1990). Experimental studies have shown that protein ubiquitination has unexpected roles in endosome–lysosome maturation (Low *et al.*, 1993), in the maturation of autophagic vacuoles (Gropper *et al.*, 1991) and in the response of some receptors to ligand binding (Cenciarelli *et al.*, 1992). Roles for protein ubiquitination in such processes could account for the presence of ubiquitin–protein conjugates in endosome–lysosomes in experimental cell systems (Doherty *et al.*, 1989a,b; Laszlo, *et al.*, 1990, 1991a,b, 1992; Low *et al.*, 1993) and for the presence of ubiquitin–protein conjugates in predominantly endosome-like structures in nerve processes in normal brain, ageing and, more particularly, in neurodegenerative diseases such as AD and LBD (Lowe *et al.*, 1993).

If we accept that APP is fragmented, at least at the carboxy-terminal truncation site, in endosome–lysosomes, the presence of ubiquitin-positive endosome–lysosomes in dystrophic neurites surrounding amyloid plaques, abutting on to some nerves cells and in the neuropil throughout the brain, is a predictable and necessary part of the pathogenesis and molecular neuropathology of Alzheimer-related neurodegenerative diseases. It is likely that the normal neuronal distribution of endosome–lysosomes becomes disrupted in those cells involved in APP fragmentation and amyloid formation, resulting in the observed distribution of ubiquitin-positive structures in diseased brain.

Confirmation of the importance of endosome–lysosomes in AD-related disorders is provided by the immunohistochemical observation of cathepsins distributed in a manner similar to that of ubiquitin–protein conjugates in or around dystrophic neurites surrounding amyloid plaques in AD. The release of cathepsins from endosome–lysosomes could result in cell and tissue damage and be responsible, at least in part, for the inflammatory reactions in and around amyloid plaques in AD.

2.5. Aβ and neurotoxicity

Demonstration of Aβ in normal brain and cerebrospinal fluid raises questions concerning possible functions or, at elevated concentrations, the neurotoxicity of such peptides. Recent evidence suggests that some aggregates of Aβ and even smaller peptides derived from the core of Aβ (e.g. residues 25–35) can be neurotoxic to primary neurones in culture (Shearman *et al.*, 1994). The mechanism of neurotoxicity is not completely understood. Currently, it is not known how small Aβ-related peptides derived from APP might cause neuronal death, but interference with calcium channels has been suggested as a mechanism. It is important to know whether Aβ-related peptides need to enter cells by endocytosis or phagocytosis in order to trigger the death process.

2.6. Apolipoprotein E4 in late-onset AD

Apolipoproteins have been known for some time to be associated with amyloid plaques in AD (Namba *et al.*, 1991). These lipoproteins are probably involved in carrying cholesterol and cholesterol esters in the brain (Mahley, 1988) which may be needed for membrane growth and repair (e.g. in processes around a developing amyloid plaque or in relation to growth cones for processes involved in new synapse formation after neuronal death in AD-related disorders). There is evidence that apolipoprotein E may be involved in regeneration after injury to the peripheral and central nervous system and in normal brain lipid metabolism (Gelman *et al.*, 1991). Apolipoprotein E binds directly to Aβ-related peptides (Strittmatter *et al.*, 1993) which may explain the presence of the lipoprotein with amyloid plaques in AD.

Recently, there have been several reports that late-onset AD is genetically associated with apolipoprotein E polymorphism, specifically copies of the apolipoprotein E4 allele (Corder *et al.*, 1993). The molecular pathological reason for this observation is still not known. However, the characteristics and locations of pathological features appear to reflect apolipoprotein isoforms (Schmechel *et al.*, 1993)

3. Transmissible neurodegenerative diseases

The prion diseases (spongiform encephalopathies) have a long history as neurodegenerative diseases of animals: scrapie has been known in sheep for some 200 years and gets its name from the propensity of animals with this disorder to rub (scrape) themselves against walls and fences, giving them a dishevelled appearance. There are several other related animal disorders, including transmissible mink encephalopathy, chronic wasting disease of mule deer and elk and, of particular current interest in the UK, BSE and a group of related disorders affecting captive rare ungulates (Prusiner, 1992).

Human patients with CJD have clinical symptoms which include a history of rapidly advancing dementia, abnormal EEG with myoclonus, other movement disorders, cortical blindness, cerebellar ataxia, rigidity and akinetic mutism (Roberts and Clinton, 1992). Similar symptoms occur in the familial autosomal dominant disorder, GSS, and in the disease kuru in Papua New Guinea, where its spread is attributed to ritualistic cannibalism. Additionally, a disorder succinctly described as fatal familial insomnia can be added to the list of prion diseases through molecular genetic characterization (Medori *et al.*, 1992) Molecular genetic techniques have also established prion disease in patients without the characteristic neuropathological features of spongiform change and PrP deposits (Collinge *et al.*, 1990).

Iatrogenic cases of CJD have occurred following treatments with pituitary-derived growth hormone or gonadotrophins and of contaminated dura mater preparations. Although the number of patients that have died from CJD is small, several thousand patients are at risk: it remains to be seen how many patients will present with clinical symptoms of prion disease. Finally, there are reports of possible transmission by blood transfusion, but these remain to be confirmed.

3.1. The PrP gene: transcription and the PrP product

PrP is coded in a single exon (*Figure 6*), excluding any possibility of alternative splicing of the primary transcript RNA to account for alternative forms of the protein. The normal form of PrP is designated PrP^C (where the superscript 'C' indicates the normal cellular form), to

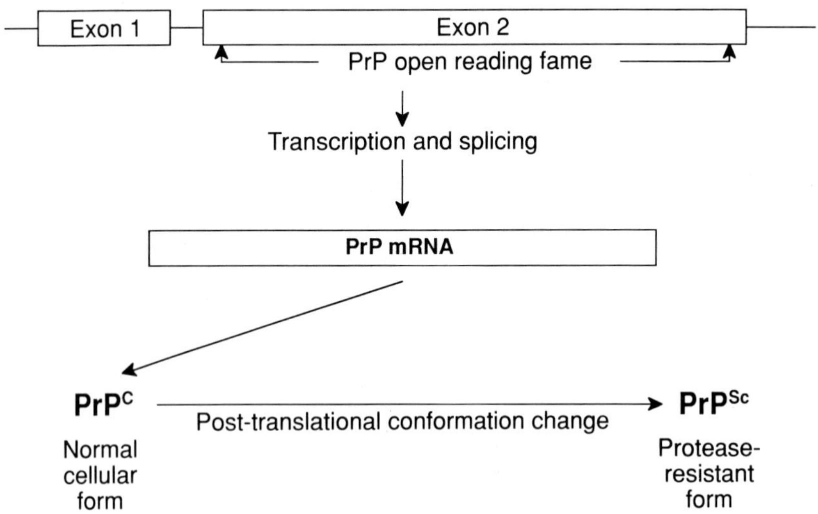

Figure 6. Transcription, translation and post-translational modification of the PrP.

distinguish it from the abnormal protein PrP^{Sc}, where 'Sc' refers to the scrapie form of the protein (or its equivalents in other species). The amino acid sequence of PrP^{C} and PrP^{Sc} is identical to the predicted amino acid sequence from the genomic DNA. No post-translational chemical modification has been identified that would distinguish PrP^{C} from PrP^{Sc}, indicating the most likely method for the formation of PrP^{Sc} is by the conformational manipulation of the normal isoform of the molecule.

PrP is a glypiated molecule anchored to the cell membrane by the phosphatidyl inositol glycan tail (*Figure 7*).

3.2. Mutations in the PrP gene

The inherited prion diseases of man are called familial CJD or GSS and are now characterized by a series of point mutations and multiple base insertions or deletions in the DNA (*Table 1*).

Table 1. Mutations in the PrP gene and relationships with prion disease

Position in sequence	Normal residue	Mutant residue	Disease	Comments or other features
53, 67, 75 or 83		Insertions	FCJD	Between four and nine octa-repeat insertions (residues 51–91 normally five octa-repeats)
81 or 82		Deletions	None evident	Single octa-repeat deletion
102	Pro	Leu	GSS	
117	Ala	Val	GSS	
129	Met	Val	Normal polymorphism	Affects susceptibility to CJD. Affects age of onset in FCJD and GSS
178	Asp	Asn	FCJD, FFI	
198	Phe	Ser	GSS	Neurofibrillary tangles
200	Glu	Lys	FCJD	
217	Glu	Arg	GSS	Neurofibrillary tangles

FCJD, familial Creutzfeldt–Jakob disease; FFI, fatal familial insomnia; GSS, Gerstmann–Straussler–Scheinker disease.

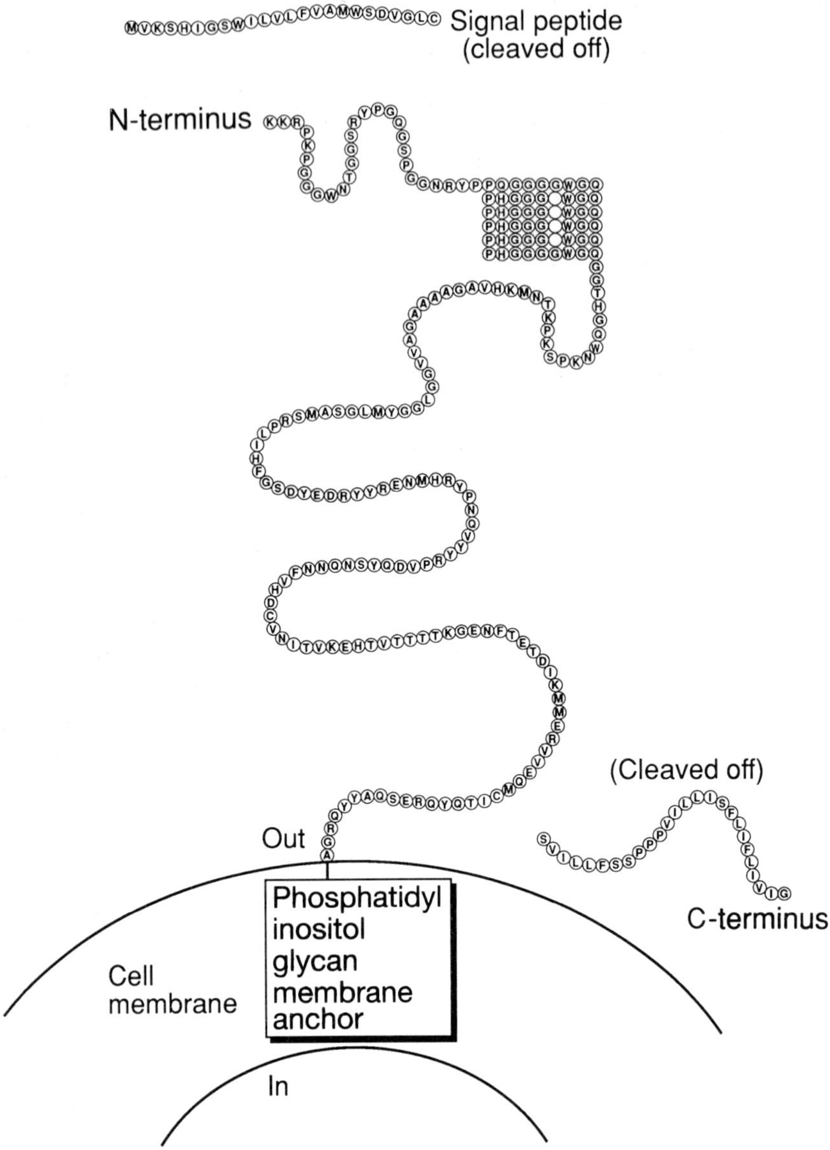

Figure 7. Primary sequence and membrane anchoring of the PrP.

The molecular genetic characterization of the PrP gene is required for every suspected case of prion disease because of varied and sometimes insignificant neuropathology.

3.3. Functions of the PrP

The functions of PrP remain unknown; a proposal that it in some way stimulated the acetyl choline receptor has since been retracted. Transgenic mice where the PrP gene has been deleted show no obvious physiological or behavioural defect (Bueler *et al.*, 1992), suggesting that its functions can be substituted by another gene product.

3.4. The infectious agent

Numerous attempts in several laboratories have failed to detect any component other than PrP in the infectious agent: there appears to be no nucleic acid (Prusiner, 1991). Transmission experiments between animal species led to the establishment of two concepts relating to infectivity; strains of agent and the species barrier. The clinical progression of disease and the neuropathological findings were different in transmission experiments with scrapie agent derived from different sources in mice of different genotypes, indicating that variants or strains of the scrapie agent occur naturally. The experimental characteristics of one strain of agent replicate faithfully on repeated infection of animals of the same genotype but not in animals of another species or, indeed, in the same species with different genotypes (Bruce *et al.*, 1991).

Such findings are pivotal to the notion of those who believe that agent variation must be attributed to a nucleic acid component with the potential for coding differences. It should be noted that the number of original 'strains' of scrapie is not large, with perhaps only three or four attributable to naturally infected sheep flocks: additional strains have been identified in laboratory mice. The seminal issue in replication of the infectious agent in transmissible encephalopathies is whether alternative forms can be generated by a protein-only mechanism. Recent findings that transgenic mice lacking the PrP gene cannot be infected with scrapie (Bueler *et al.*, 1993) highlight the issue of the key role of host PrP for agent replication. Questions of how and where prions in the infectious agent can interact with host PrP^C and generate more infectious particles become ever more pertinent.

3.5. The conversion of PrP^C to PrP^{Sc}

Inability to detect covalent chemical differences between PrP^{Sc} in agent preparations and PrP^C isolated from brain has been interpreted in terms of an alteration in the conformation of PrP^C being induced by PrP^{Sc} when infection occurs (Prusiner, 1992). Host PrP^C, internalized from the cell membrane by some form of endocytosis, will enter the endosome–lysosome system for degradation to amino acids. Presumably, as with APP, the degradation of the protein is a complex multi-step process involving unfolding and dissolution by proteases and other hydrolases. Although not characterized, a thermodynamically determined set of reaction steps with defined low concentrations of intermediates will be part of the degradation process. After experimental infection of animals by direct injection into the brain, prions will be taken up by phagocytosis into a variety of cells, including neurones. In this way prions would be introduced directly into the endosome–lysosome system: a critically important fact relevant to the replication of a protein-only infectious agent. Within the cisterns of the endosome–lysosome system, PrP^{C}, entering normally for catabolism, will be unfolded, thus exposing protein surfaces which, through secondary structural interactions, can form the intermolecular β-sheet arrangements that are the basis of amyloid formation. The slow but continuous import of PrP^{C} into the endosome–lysosome system will result in the accumulation of PrP^{Sc}, which may either be expelled from the cell following fusion of endosome–lysosomes with the cell membrane or may accumulate to the point at which the organelle bursts, releasing prions and hydrolases, many of which are still active at neutral pH, into the cytoplasm. The hydrolases will cause cytolysis, leading to cell death, spongiform change and host morbidity (*Figure 8*).

In experimental infection of rodents, these events take times varying from around 60 to 550 days or longer, depending both on species and strain of the agent. In the familial prion diseases, mutations in the PrP gene will render degradation of the protein within the endosome–lysosome more difficult, compromising the dissolution pathway so that higher levels of unfolding or protein fragment intermediates are generated that are capable of intermolecular secondary structural interactions to form PrP^{Sc}. There is evidence that fragments of PrP occur in preparations of prions (Tagliavini *et al.*, 1991): it has also been shown that patients with a stop codon in the open reading frame of the PrP^{C} gene on one allele exhibit the clinical features of CJD and produce amyloid based on the truncated PrP molecules (Kitamoto *et al.*, 1993). There is also the possibility that fragments of PrP may play a central role in the conversion of PrP^{C} into PrP^{Sc} by acting to nucleate the process.

Immunogold electron microscope studies on scrapie-infected mice have shown that PrP^{Sc} accumulates together with ubiquitin–protein conjugates, an hsp 70, β-glucuronidase and the cation-independent mannose-6-phosphate receptor in endosome–lysosomes: the latter observation suggests that late multivesicular endosomes have a key role

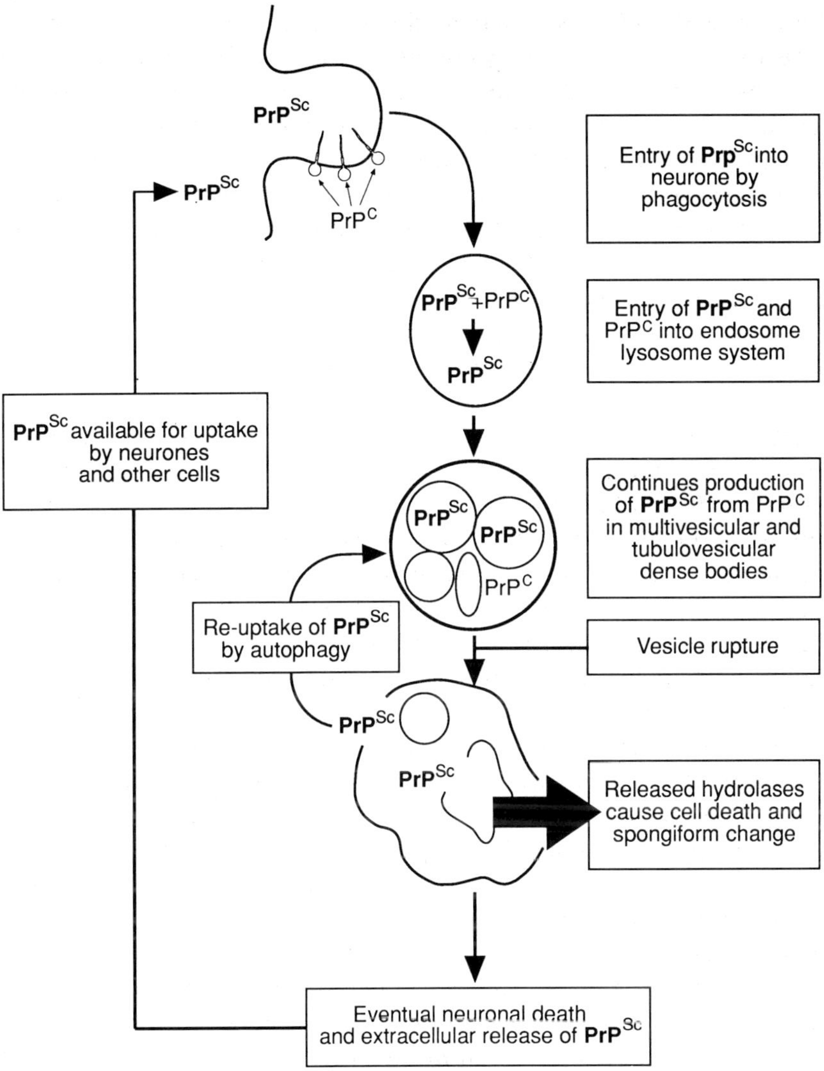

Figure 8. Proposed role of endosome–lysosomes in the production of PrP^{Sc} and associated pathological changes in prion diseases.

in the conversion of PrP^{C} into PrP^{Sc} (Arnold, Tipler, Hope, Landon, Laszlo and Mayer, unpublished observations). PrP^{Sc} similarly accumulates in this family of organelles in scrapie-infected neuroblastoma cells (Borchelt *et al.*, 1992). As the disease draws to its fatal conclusion, electron microscopic evidence for intraneuronal spongiform lesions becomes apparent, with early spongiform change visible adjacent to dense endosome–lysosome-like multivesicular and tubulovesicular bodies (Lowe *et al.*, 1992). Observations on scrapie-infected mouse brain and neuroblastoma cells support the notion that endosome–lysosomes have a central role in the biogenesis of prions: the organelles act as bioreactors for the abnormal isoform of PrP (Arnold, Tipler, Hope, Landon, Laszlo and Mayer, unpublished observations).

Generation of protein-only infectious particles in endosome–lysosomes can account for strains of scrapie-like agents without the need for nucleic acids. The degradation of proteins in endosome–lysosomes will be thermodynamically influenced through the stability of intermediates and kinetically by the rates of their interconversions. The denaturing environment of the organelles, together with the complex mixture of proteases and other hydrolytic enzymes, means that there may be alternative pathways for the unfolding and degradation of a single protein. The unfolding and hydrolytic pathways are likely to vary between cell types, depending on cell function (cf. microglial cells and neurones) and, undoubtedly, will be influenced by the genotype of the infected animal. Given that different fragments of PrP can be produced, then alternative products may nucleate the formation of varying types of PrP^{Sc}-containing amyloid by triggering one or other of a range of secondary structural interactions and intermolecular β-sheet conformations. This type of process could give rise to a thermodynamically-controlled, limited number of alternative strains of amyloids, which can faithfully reproduce themselves on reinfection of new animals of the same genotype (DeArmond, 1993).

Prion replication in the endosome–lysosome system of cells also explains why cells of the reticulo-endothelial system produce infectious agent (Fraser *et al.*, 1992). Some of the cells in the reticulo-endothelial system are antigen-processing cells, with endosome–lysosomes designed to fragment antigens for presentation in association with class II histocompatibility molecules: such organelles are likely to generate PrP^{Sc} from PrP^{C} in the same way as has been suggested for other cell types in the nervous system.

3.6. Fragments of PrP and neurotoxicity

Fragments of PrP have been shown to be neurotoxic *in vitro*, where they induce neuronal death by a nucleolytic endonuclease-mediated process (Forloni *et al.*, 1993). Fragments of highest neurotoxicity correspond to a region (residues 106–126) which has a propensity to form aggregates spontaneously and a tendency to form β-pleated sheets *in vitro*. It is conceivable that some fragments of PrP have the capacity to form amyloids and concurrently be neurotoxic, which would contribute to disease progression and the development of neuropathology. The *in vitro* demonstration of neuronal death by DNA fragmentation induced by potentially amyloidogenic sequences of PrP may not occur *in vivo*. Attempts to measure DNA fragmentation in brain regions susceptible to neurodegeneration and cell death have given negative results (Mighelli, Landon, Lowe, Fergusson and Mayer, unpublished observations), indicating that there may be several mechanisms of cell death. Interestingly, two proteolysis-based mechanisms have emerged recently in studies on models of programmed neuromuscular cell death. One mechanism involves ubiquitin-dependent proteolysis (Schwartz *et al.*, 1990), the other the intracellular cysteine protease, interleukin-converting enzyme (Yuan *et al.*, 1993). The increased presence of endosome–lysosomes in nerve processes and the elevated level of polyubiquitin gene expression in the terminal stages of prion disease (Kenward *et al.*, 1993) might indicate a role for ubiquitin-dependent autophagic degradation of the cytoplasm in such disorders (and in AD-related diseases). The great advantage of autophagy for programmed cell death is that cell corpses are generated which may be simply removed by phagocytosis, thus minimizing or avoiding immune involvement in the brain.

4. Endosome–lysosome bioreactors: a commonality in neurodegenerative diseases

The combined studies on idiopathic and transmissible neurodegenerative diseases present observations which allocate a key role to the post-functional processing of single membrane proteins within the endosome–lysosome system in the generation of amyloidogenic protein aggregates.

Single membrane protein and related disorders can be of two types; those in which genetic abnormalities are manifested by biosynthetic and functional abnormality (e.g. cystic fibrosis and muscular dystrophy) and those diseases in which the post-functional catabolism of a protein may generate pathological protein-based materials (e.g. AD and prion diseases). It could be argued that there is no cellular site, other than the interior of an endosome–lysosome, that has the capacity to catalyse the protein–protein interactions which eventually result in the genesis of amyloid (Arnold, Tipler, Hope, Landon, Laszlo and Mayer, unpublished observations). The organelles are designed to unfold and degrade macromolecules and, therefore, are a site at which, over many years, the dismantling of membrane proteins could generate intermediates for amyloid formation. There is an exponential-like element to disease progression and neuropathology in transmissible neurodegenerative diseases. Such an exponential process would be predicted on the basis of repeated rounds of phagocytosis of prions, endosome–lysosome replication and release of prions during cell death. Similar events could generate amyloid in AD.

5. Gene therapy

The complexities of neurodegenerative diseases, particularly the involvement of a complex organelle system in amyloidogenesis, makes single gene therapy difficult to anticipate. The obvious approach in the long term would be deletion of the genes for APP and PrP from the genome, if their normal functions finally are conclusively proven to be non-essential. This is currently a serious proposition for farm animals, including cattle, in relation to the epidemic of BSE. Ethical considerations aside, it will of course be possible in the not too distant future to delete the APP and the PrP genes from the human genome, thus obviating the diseases, always assuming that amyloidogenesis is the key factor in molecular pathogenesis.

References

Benowitz LI, Rodriguez W, Paskevich P, Mufson EJ, Schenk D and Neve RL. (1989) The amyloid precursor protein is concentrated in neuronal lysosomes in normal and Alzheimer disease subjects. *Exp. Neurol.* **106**, 237–250.

Borchelt DR, Taraboulos A and Prusiner SB. (1992) Evidence for synthesis of scrapie prion proteins in the endocytic pathway. *J. Biol. Chem.* **267**, 16188–16199.

Brion JP, Hanger DP, Bruce MT, Couck AM, Flamentdurand J and Anderton BH. (1991) Tau in Alzheimer neurofibrillary tangles – N-terminal and C-terminal regions are differentially associated with paired helical filaments and the location of a putative abnormal phosphorylation site. *Biochem. J.* **273**, 127–133.

Bruce ME, McConnell I, Fraser H and Dickinson AG. (1991) The disease characteristics of different strains of scrapie in *Sinc* congenic mouse lines: implications for the nature of the agent and host control of pathogenesis. *J. Gen. Virol.* **72**, 595–603.

Bueler H, Fischer M, Lang Y, Bluethmann H, Lipp H-P, DeArmond SJ, Prusiner SB, Aguet M and Weissmann C. (1992) Normal development and behaviour of mice lacking the neuronal cell-surface PrP protein. *Nature,* **356**, 577–582.

Bueler H, Aguzzi A, Sailer A, Greiner R-A, Autenreid P, Aguet M and Weissmann C. (1993) Mice devoid of PrP are resistant to scrapie. *Cell,* **73**, 1339–1347.

Byrne EJ, Lennox GG, Godwin-Austen RB, Jeffereson D, Lowe J, Mayer RJ, Landon M and Doherty FJ. (1991) Dementia associated with cortical Lewy bodies: proposed clinical diagnostic criteria. *Dementia,* **2**, 283–284.

Cenciarelli C, Hou D, Hsu KC, Rellahan BL, Wiest DL, Smith HT, Fried VA and Weissman AM. (1992) Activation-induced ubiquitination of the T-cell antigen receptor. *Science,* **257**, 795–797.

Collinge J, Owen F, Poulter M *et al.* (1990) Prion dementia without characteristic pathology. *Lancet,* **336**, 7–9.

Corder EH, Saunders AM, Strittmatter WJ, Schmechel DE, Gaskell PC, Small GW, Roses AD, Haines JL and Pericak-Vance MA. (1993) Gene dose of apolipoprotein E type 4 allele and the risk of Alzheimer's disease in late onset families. *Science,* **261**, 921–923.

Davis JQ, Dansereau D, Johnstone RM and Bennett V. (1986) Selective externalization of an ATP-binding protein structurally related to clathrin uncoating ATPase heat shock protein in vesicles containing terminal transferrin receptors during reticulocyte maturation. *J. Biol. Chem.* **261**, 15368–15371.

DeArmond SJ. (1993) Alzheimer's disease and Creutzfeldt–Jakob disease: overlap of pathogenic mechanisms. *Curr. Opin. Neurol.* **6**, 872–881.

Dickson DW, Wertkin A, Kress Y, Ksiezak-Reding H and Yen S-H. (1990) Ubiquitin immunoreactive structures in normal human brains. *Lab. Invest.* **63**, 87–99.

Doherty FJ, Osborn NU, Wassell JA, Laszlo L and Mayer RJ. (1989a) Insoluble disulphide cross-linked polypeptides accumulate in the functionally compromised lysosomes of fibroblasts treated with the cysteine protease inhibitor E-64. *Exp. Cell Res.* **185**, 506–518.

Doherty FJ, Osborn NU, Wassell JA, Laszlo L and Mayer RJ. (1989b) Cysteine protease inhibitors cause the accumulation of ubiquitin-polypeptide conjugates in lysosomes of 3T3-L1 fibroblasts. *Biochem. Soc. Trans.* **17**, 171–172.

Earl RT, Mangiapane EH, Billett EE and Mayer RJ. (1987) A putative protein-sequestration site involving intermediate filaments for protein degradation by autophagy. *Biochem. J.* **241**, 809–815.

Esch FS, Keim PS, Beattie EC, Blacher RW, Culwell AR, Oltersdorf T, McClure D and Ward PJ. (1990) Cleavage of amyloid β-peptide during constitutive processing of its precursor. *Science,* **248**, 1122–1124.

Forloni G, Angeretti N, Chiesa R, Monzani E, Salmona M, Bugiani O and Tagliavini F. (1993) Neurotoxicity of a prion protein fragment. *Nature,* **362**, 543–546.

Fraser H, Bruce ME, Davies D, Farquhar CF and McBride PA. (1992) The lymphoreticular system in the pathogenesis of scrapie. In: *Prion Diseases of Humans and Animals* (eds SB Prusiner, J Collinge, J Powell and B Anderton). Ellis Horwood, London, pp. 308–317.

Gelman BB, Goodrum J and Bouldin TW. (1991) Macrophage apolipoprotein synthesis and endoneural distribution as a response to segmental demyelination. *J. Neuropathol. Exp. Neurol.* **50**, 383–407.

Gropper R, Brandt RA, Elias S, Bearer CF, Mayer A, Schwartz AL and Ciechanover A. (1991) The ubiquitin-activating enzyme, E1, is required for stress-induced lysosomal degradation of cellular proteins. *J. Biol. Chem.* **266**, 3602–3610.

Haass C, Koo EH, Mellon A, Hung AY and Selkoe DJ. (1992) Targeting of cell-surface β-amyloid precursor protein to lysosomes: alternative processing into amyloid-bearing fragments. *Nature,* **357**, 500–503.

Hansen SL, Masliah E, Terry RD and Mirva SS. (1989) A neuropathological subset of Alzheimer's disease with concomitant Lewy body disease and spongiform change. *Acta Neuropathol.* **78** 194–201.

Hansen L, Salmon D, Galasko D and Masliah E. (1990) The Lewy body variant of Alzheimer's disease: a clinical and pathological entity. *Neurology,* **40**, 1–8.

Jarrett JT, Berger EP and Lansbury PT. (1993) The carboxy terminus of the beta-amyloid protein is critical for the seeding of amyloid formation: implications for the pathogenesis of Alzheimer's disease. *Biochemistry,* **32**, 4693–4697.

Kenward N, Hope J, Landon M and Mayer RJ. (1994) Expression of polyubiquitin and heat-shock protein 70 genes increases in the later stages of disease progression in scrapie-infected mouse brains. *J. Neurochem,* in press.

Kitamoto T, Iizuka R and Tateishi J. (1993) An amber mutation of prion protein in Gerstmann–Straussler syndrome with mutant PrP plaques. *Biochem. Biophys. Res. Commun.* **192**, 525–531.

Konig G, Monning U, Czech C, Prior R, Banati R, Schreiter-Gasser U, Bauer J, Masters CL and Beyreuther K. (1992) Identification and differential expression of a novel alternative splice isoform of the βA4 amyloid precursor protein (APP) mRNA in leukocytes and brain microglial cells. *J. Biol. Chem.* **267**, 10804–10809.

Laszlo L, Doherty FJ, Osborn NU and Mayer RJ. (1990) Ubiquitinated protein conjugates are specifically enriched in the lysosomal system of fibroblasts. *FEBS Lett.* **261**, 365–368.

Laszlo L, Doherty FJ, Watson A, Self T, Landon M, Lowe J and Mayer RJ. (1991a) Immunogold localisation of ubiquitin–protein conjugates in primary (azurophilic) granules of polymorphonuclear neutrophils. *FEBS Lett.* **279**, 175–178.

Laszlo L, Tuckwell J, Self T, Lowe J, Landon M, Smith S, Hawthorne JN and Mayer RJ. (1991b) The latent membrane protein-1 in Epstein–Barr virus-transformed lymphoblastoid cells is found with ubiquitin–protein conjugates and heat-shock protein-70 in lysosomes oriented around the microtubule organizing centre. *J. Pathol.* **164**, 203–214.

Laszlo L, Lowe J, Self T *et al.* (1992) Lysosomes are key organelles in the pathogensis of prion encephalopathies. *J. Pathol.* **166**, 333–341.

Lennox G, Lowe J, Landon M, Byrne EJ, Mayer RJ and Godwin-Austen RB. (1989a) Diffuse Lewy body disease: correlative neuropathology using anti-ubiquitin immunocytochemistry. *J. Neurol. Neurosurg. Psychiatr.* **52**, 1236–1247.

Lennox G, Lowe J, Morrell K, Landon M and Mayer RJ. (1989b) Anti-ubiquitin immunocytochemistry is more sensitive than conventional techniques in the detection of diffuse Lewy body disease. *J. Neurol. Neurosurg. Psychiatr.* **52**, 67–71.

Low P, Doherty FJ, Sass M, Kovacs J, Mayer RJ and Laszlo L. (1993) Immunogold localisation of ubiquitin–protein conjugates in Sf9 insect cells. Implications for the biogenesis of lysosome-related organelles. *FEBS Lett.* **316**, 152–156.

Lowe J and Mayer RJ. (1990) Ubiquitin, cell stress and diseases of the nervous system. *Neuropathol. Appl. Neurobiol.* **16**, 281–291.

Lowe J, Fergusson J, Kenward N, Laszlo L, Landon M, Farquhar C, Brown J, Hope J and Mayer RJ. (1992) Immunoreactivity to ubiquitin–protein conjugates is present early in the disease process in the brains of scrapie-infected mice. *J. Pathol.* **168**, 169–177.

Lowe J, Mayer RJ and Landon M. (1993) Ubiquitin in neurodegenerative diseases. *Brain Pathol.* **3**, 55–65.

Mahley RW. (1988) Apolipoprotein E: cholesterol transport protein with expanding role in cell biology. *Science,* **240**, 622–629.

Mayer RJ, Arnold J, Laszlo L, Landon M and Lowe J. (1991a) Ubiquitin in health and disease. *Biochim. Biophys. Acta,* **1089**, 141–157.

Mayer RJ, Lowe J and Landon M. (1991b) Ubiquitin and the molecular pathology of chronic degenerative diseases. *J. Pathol.* **163**, 279–281.

Mayer RJ, Landon M, Laszlo L, Lennox G and Lowe J. (1992) Protein processing in lysosomes – the new therapeutic target in neurodegenerative disease. *Lancet,* **340**, 156–159.

McKinley MP, Taraboulos A, Kenaga L, Serban D, Stieber A, DeArmond SJ, Prusiner SB and Gonatas N. (1991) Ultrastructural localisation of scrapie prions in cytoplasmic vesicles of infected cultured cells. *Lab. Invest.* **65**, 622–630.

Medori R, Tritschler H-J, Le Blanc A, Villare F, Manetto V and Chen HY. (1992) Fatal familial insomnia, a prion disease with a mutation at codon 178 of the prion protein gene. *N. Engl. J. Med.* **326**, 444–449.

Namba Y, Tomonaga M, Kawasaki H, Otomo E and Ikeda K. (1991) Apolipoprotein E immunoreactivity in cerebral amyloid deposits and neurofibrillary tangles in Alzheimer's disease and Kuru plaque amyloid in Creutzfeldt–Jakob disease. *Brain Res.* **541**, 163–166.

Okazaki H, Lipkin LE and Aronson SM. (1961) Diffuse intracytoplasmic ganglionic inclusions (Lewy body type) associated with progressive dementia and quadriparesis in flexion. *J. Neuropathol. Exp. Neurol.* **20**, 237–244.

Oltersdorf T, Fritz LC, Schenk DB *et al.* (1989) The secreted form of the Alzheimer's amyloid precursor protein with the Kunitz domain is protease nexin III. *Nature,* **341**, 144–147.

Parton RG and Dotti CG. (1993) Cell biology of neuronal endocytosis. *J. Neurosci. Res.* **36**, 1–9.

Parton RG, Simons K and Dotti CG. (1992) Axonal and dendritic endocytic pathways in cultured neurons. *J. Cell Biol.* **119**, 123–137.

Perry RH, Irving D, Blessed G, Perry EK and Fairburn AF. (1989) Senile dementia of Lewy body type and spectrum of Lewy body disease. *Lancet,* **8646**, 1088.

Price DL, Borchelt DR and Sisodia SS. (1993) Alzheimer disease and the prion disorders amyloid β-protein and prion protein amyloidoses. *Proc. Natl Acad. Sci USA,* **90**, 6381–6384.

Prusiner SB. (1991) Molecular biology of prion diseases. *Science,* **252**, 1515–1522.

Prusiner SB. (1992) Chemistry and biology of prions. *Biochemistry,* **31**, 12277–12288.

Roberts GW and Clinton J. (1992) Prion disease: the spectrum of pathology and diagnostic considerations. In: *Prion Diseases of Humans and Animals* (eds SB Prusiner, J Collinge, J Powell and B Anderton). Ellis Horwood, London, pp. 225–242.

Roberts GW, Gentleman SM, Lynch A, Murray L, Landon M and Graham DI. (1994) β-Amyloid protein deposition in the brain after severe head injury: implications for the pathogenesis of Alzheimer's disease. *J. Neurol. Neurosurg. Psychiatr,* in press.

Roher AS, Lowenson JD, Clarke S, Woods AS, Cotter RJ, Gowing E and Ball MJ. (1993a) β-Amyloid 1–42 is a major component of cerebrovascular amyloid deposits: implications for the pathology of Alzheimer's disease. *Proc. Natl Acad. Sci. USA,* **90**, 10836–10840.

Roher AE, Palmer KC, Yurewicz EC, Ball MJ and Greenburg BD. (1993b) Morphological and biochemical analyses of amyloid plaque core proteins purified from Alzheimer's disease brain tissue. *J. Neurochem.* **61**, 1916–1926.

Schmechel DE, Saunders AM, Strittmatter WJ, Crain BJ, Hulette CM, Joo SH, Pericak-Vance MA, Goldgaber D and Roses AD. (1993) Increased amyloid β-peptide deposition in cerebral cortex as a consequence of apolipoprotein E genotype in late onset Alzheimer's disease. *Proc. Natl Acad. Sci. USA,* **90**, 9649–9653.

Schmidt ML, Murray J, Lee VM, Hill WD, Wertkin A and Trojanowski JQ. (1991) Epitope map of neurofilament protein domains in cortical and peripheral nervous system Lewy bodies. *Am. J. Pathol.* **139**, 53–65.

Schwartz LM, Myer A, Kosz L, Engelstein M and Maier C. (1990) Activation of polyubiquitin gene expression during developmentally programmed cell death. *Neuron,* **5**, 411–419.

Selkoe DJ. (1991) The molecular pathology of Alzheimer's disease. *Neuron,* **6**, 487–498.

Seubert P, Oltersdorf T, Lee MG *et al.* (1993) Secretion of β-amyloid precursor protein cleaved at the amino terminus of the β-amyloid peptide. *Nature,* **361**, 260–263.

Shearman MS, Ragan CI and Iversen LL. (1994) Inhibition of PC12 cell redox activity is a specific, early indicator of the mechanism of β-amyloid-mediated cell death. *Proc. Natl Acad. Sci. USA,* in press.

Siman R, Mistretta S, Durkin JT, Savage MJ, Loh T, Trusko S and Scott

RW. (1993) Processing of the β-amyloid precursor. *J. Biol. Chem.* **268**, 16602–16609.

Strittmatter WJ, Weisgraber KH, Huang DY *et al.* (1993) Binding of human apolipoprotein E to synthetic amyloid β-peptide: isoform specific effects and implications for late onset Alzheimer's disease. *Proc. Natl Acad. Sci. USA*, **90**, 8098–8102.

Tagliavini F, Prelli F, Ghiso J, Bugiani O, Serban D, Prusiner SB, Ghetti B and Frangione B. (1991) Amyloid protein of Gerstmann–Straussler–Scheinker disease (Indiana kindred) is an 11 kd fragment of prion protein with an N-terminal glycine at codon 58. *EMBO J.* **10**, 513–519.

Yuan J, Shamam S, Ledoux S, Ellis HM and Horvitz HR. (1993) The *C. elegans* cell death gene *ced-3* encodes a protein similar to mammalian interleukin Iβ converting enzyme. *Cell*, **75**, 641–652.

13

Herpes simplex: once bitten, forever smitten?

D.S. Latchman
Department of Molecular Pathology, University College London Medical School, London, UK

1. Introduction

The fundamental dogma of molecular biology is that the genetic material in the DNA which is inherited from your parents is then converted into RNA and that RNA makes the proteins that are characteristic of each tissue, be it globin in red blood cells, myosin in muscle, etc. One question of great interest is why, if the genetic material is the same in each cell, does each cell make a different set of proteins? For example, why does the muscle make myosin, the red blood cell make globin and so on, even though the genes are present and effectively the same in all different tissues.

Working in conjunction with Martin Evans in Cambridge some years ago, where we were looking at the gene for alphafetoprotein, a protein which is made only in the fetal liver, we rapidly realized that this was going to be very difficult, simply because of the vast size of the human genome; human DNA consisting of 3000 million base pairs in each individual cell. This means that even if the individual gene can be extracted, isolated and characterized, it is still interacting with a vast amount of other DNA. Therefore it was decided to move to some form of simpler model system. Many people have studied bacteria, which are about 1000 times simpler in terms of their DNA (2×10^6 base pairs of DNA per *Escherichia coli* cell) and much easier to study. The problem, of course, is that bacteria do not have muscles, blood cells, a liver or a brain, so obviously the problem of tissue-specific gene expression cannot be studied in bacteria. The simpler you go, the easier it is but the less you can find out.

So what is the answer to this? Well, the answer, paradoxically, is to go even simpler: go down to the virus, where the genome size is even smaller. For example, as when working on the SV40 virus, with Peter Rigby at Imperial College, there is very much less DNA (5×10^3 base pairs) and they are much simpler, even than *E. coli,* and certainly than humans. It could be said, and quite rightly, that if viruses are so much simpler then surely even less can be learned by studying viruses than can be learned by studying bacteria. The answer to this question is shown in *Figure 1*, which illustrates the distinction between a bacterium and a virus. A bacterium is essentially an independent organism, it may secrete toxins etc., which influence the host cell, but the process in which its DNA makes RNA makes protein is taking place in an independent capsule from the host cell. In contrast, the virus, because it is so much simpler, cannot do that on its own, all it can do is to go into

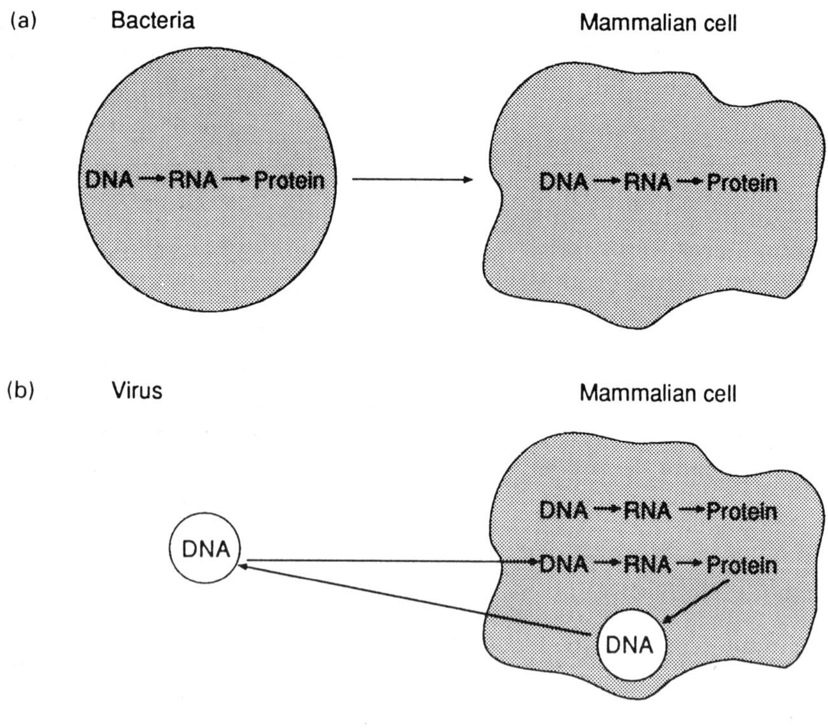

Figure 1. Interaction of (a) bacteria and (b) viruses with mammalian cells.

the cell with its DNA or RNA and use the cellular machinery to make its proteins and to copy itself. Because of this, the virus may be used as a probe to actually get into the cell and, once there, help the researcher to look at what is going on within the cell, because this simple thing with its very small amount of DNA is actually going in there and elicting information which can be used to look at the cell.

2. Herpes simplex virus

Work is being carried out at UCL on herpes simplex virus (HSV), which infects the majority of the human population. Something like 70% of people have been exposed to this virus by late adolescence and by the time they reach old age, or relative old age, virtually everybody has been infected. This virus causes facial cold sores, which are experienced by something like 40% of the population at one time or another.

It is not surprising that something as ubiquitous and as well known as herpes has been known right back to ancient times. Indeed, Hippocrates used the term 'herpes', which is the Greek for 'creeping', to describe the spreading nature of these lesions. They were therefore recognized extremely early and they were studied but, as so often happens in medical science, nothing actually emerged about these viruses until the end of the 19th century. The first step forward was the work of Vidal, who observed in 1893 that herpes infections were transmitted from one individual to another. Vidal's studies were epidemiological, in that he noticed that there was a pattern of infectious spread in agreement with there being an infectious agent. Much later, in 1919, Lowenstein did the direct experiment of taking material from human lesions and producing similar lesions when he innoculated the material into the rabbit eye. This, of course, is the ultimate test for an aetiological agent, namely that it can be transferred to another organism. A year later, Gruter reported the results of experiments carried out 8 years before in which he took the material from the rabbits and innoculated it into the eye of a blind man to produce the same thing.

2.1. HSV latency

What is clear from these experiments, therefore, is that it was established by 1920 at least that there was an infectious agent and that this infectious agent was a virus. However, at this time there were still a great many problems in this whole field, even though the agent was established. These problems arose from the fact that the facial cold sores caused by HSV recur and that they recur at the same place on the mouth each time. Moreover, they recur in the absence of any new exposure to the virus, with some people reactivating the virus every few months and others much more rarely or not at all. So that, as described earlier, although virtually everybody has the virus, only about 40% of people experience cold sores at all.

Indeed, as late as 1938, Doerr was denying the infectious nature of herpes lesions because, as he said, these sores recur repeatedly without new infection, so they must be generated endogenously. This idea actually harks back to the Roman physician Galen who said that herpes lesions were an attempt by the body to rid itself of evil humours and he called it 'herpes excretins' for this reason. There was therefore a great dispute going on in the 1920s and 1930s between those who had actually established that there was a virus and those who said this could not be the case because the cold sore keeps recurring with no new infection. Interestingly, Shakespeare had his own interpretation of this and, in Romeo and Juliet (Act I, Scene IV), he describes the activities of the fairies' midwife Mab, and says:

> 'And in this state she gallops night by night . . . O'er ladies lips, who straight on kisses dream which oft the angry Mab with blisters plagues'.

So, that was his solution to the problem and would that it were so simple and we could give up at this point and put it all down to the fairies!

In fact, the real solution to this problem was provided in about 1905 by the neurosurgeon Harvey Cushing. Harvey Cushing was the identifier of Cushing's Disease, caused by overactivity of the adrenal gland which is still, of course, observed today. However, he was also a pioneer of neurosurgery and was one of the first to conduct neurosurgery for trigeminal neuralgia, which results in acute pain due to abnormal sensory nerve activity. In 1905 he reported the results of his experiments involving removing or cutting the trigeminal nerve (Cushing, 1905). Obviously, if the sensory nerve is cut then the pain is relieved.

In addition, however, Cushing made a very interesting observation. He saw in some patients that there was an area of anaesthesia in which the patient is relieved from pain by the cutting of the sensory nerve and in the region immediately adjacent the eruption of herpes sores is seen. He made this observation on several patients with sores appearing, not in the area of anaesthesia, but in the area immediately adjacent. In a brilliant jump forward, he reasoned that the only way in which these two could be connected was through the nerve cell bodies in the ganglia so that if a cut is made in a particular branch of the nerve and then the virus is seen emerging at another point, there must be some nervous involvement; it is not simply a matter of peripheral lesions (*Figure* 2).

On the basis of many other experiments since, it is now known that at the time of the first lytic infection, where the virus is growing, replicating and producing the cold sore peripherally, it makes its way up the processes of the sensory nerve innervating that tissue, and then hides itself away in the sensory nerve cell bodies (*Figure* 3). When the peripheral infection has been cleared up by drugs, as will be described later, or by the immune system, the infection persists in the nerve cells in a silent form which is called a 'latent infection'. However, in response to certain stimuli the virus re-emerges, comes back down and, when it finds the susceptible epithelial cells again, starts a new lytic infection. This is why a new recurrence of the cold sore is seen and it is at the same place, of course, because it is the same nerve along which the virus is travelling (for reviews, see Latchman, 1990; Roizman and Sears, 1987; Stevens, 1975).

In terms of gene regulation, this is a really brilliant system because,

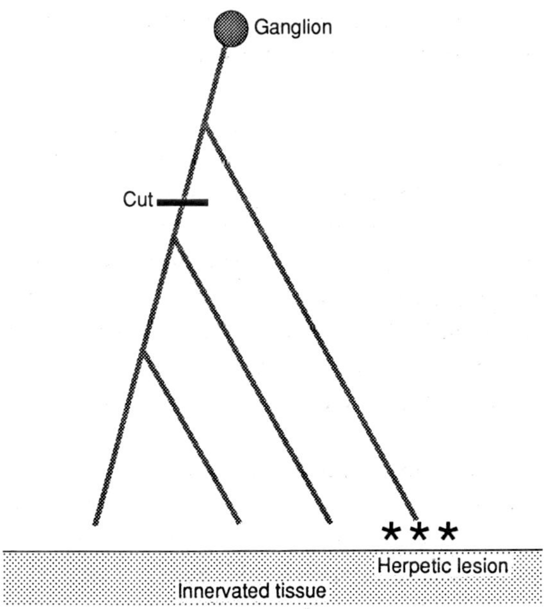

Figure 2. Cutting of the nerve process results in herpetic lesions at sites adjacent to the region innervated by the cut nerve.

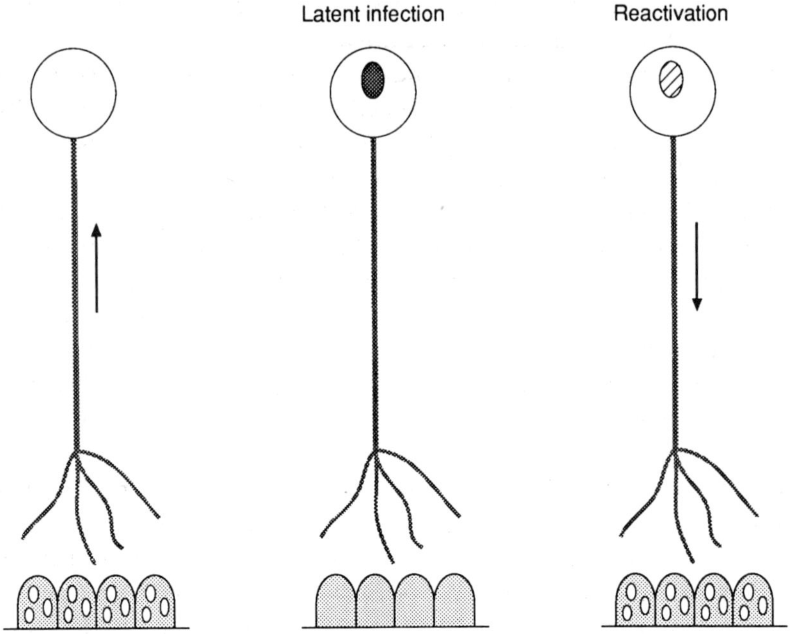

Figure 3. The stages of HSV infection. Initial lytic infection of epithelial cells is followed by the establishment of asymptomatic latent infections which can subsequently reactivate, leading to new lytic infection.

in situations where the virus can express itself, such as in the epithelial cell type, its proteins are made, whereas in the neuron it does not express its proteins. So the virus is the same in both cases and yet its action is completely different. That must be a function of the cell itself: it must be that the nerve cell has a different complement of factors which allow it to restrict infection where the epithelial cell cannot do so. This therefore provides a beautiful model of gene control and, of course, it gets better than that because many of the stimuli which reactivate actually change the state of the nerve. Thus, changing the state of the nerve can reactivate the virus and bring it back down again. For example, cutting the nerve process, as in the Cushing experiment, leads to some change which results in the virus reactivating and coming back down the nerve and producing a lytic infection (*Figure 4*). Other stimuli that can do this are emotional factors, so again changes in the nerve lead to reactivation. This indicates, therefore, that the virus may be used as a probe for the tissue-specific differences in gene expression between cell types and as a sensor for the state of the nerve.

This, of course, is sufficient to make the virus extremely interesting but what makes it even more interesting is that this is a clinically important phenomenon as well. Thus, if a patient visits the doctor with a cold sore they usually will be treated with the Wellcome drug Acyclovir (Zovirax). Acyclovir is very effective and it clears up the sore because it acts to inhibit the virus's DNA replication. So, the patient goes away happy, with no side-effects at all. What happens then, however, is that some time later in response to any of these stimuli, the virus comes back again because Acyclovir has no effect on a virus which is in a silent form in the nerve cells. Therefore, latency is the clinically important phenomenon because lytic infection can be treated but latency remains unaffected.

It could be said that it is a bit distressing to have cold sores but it is not really life threatening, we all have other things to worry about – AIDS, heart disease and so on – why should we bother about herpes cold sores? Well, the virus does actually have lethal effects as well. In the newborn for example, the virus is lethal if the baby is infected by the

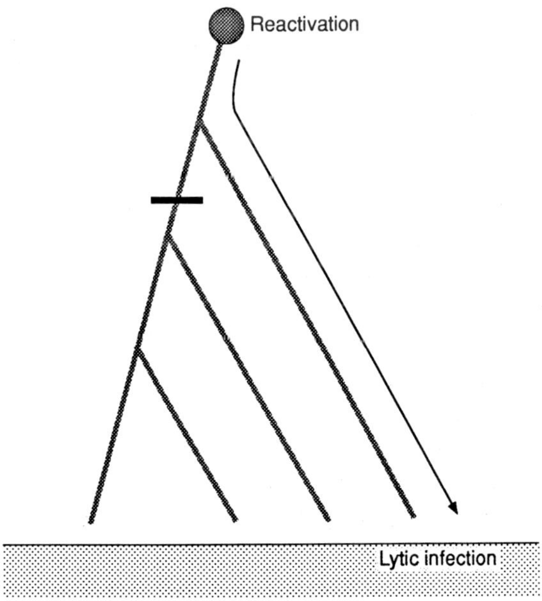

Figure 4. The appearance of herpetic lesions adjacent to the area innervated by the cut nerve is explained by the reactivation of virus in the nerve cell bodies.

mother, with 25% mortality. Similarly, in immuno-compromised people such as transplant patients or AIDS patients, the virus emerges from latency and they cannot cope with the disease. Finally, there is herpes encephalitis infection of the brain which is very rare but is almost invariably lethal or results in severe brain damage. All these cases actually illustrate the importance of reactivation of latent virus. Thus, what is going on in the transplant patient or AIDS patient is not a new infection but a reactivation of latent virus. So the fact that latent virus is commonly present is dangerous if someone becomes immuno-compromised. Similarly, it is dangerous for people if it reactivates in the brain and leads to encephalitis. Therefore, in order to respond to the problem of herpes, the problem of latent infection must be addressed.

It becomes clear that HSV latency is interesting in terms of gene regulation and in terms of the clinical phenomenon as well. In lytic infection in most cells the virus can make its proteins, can replicate its DNA and kill the cell while, in contrast, in neuronal cells it does not do so. Studies will be described here which have attempted to explain why that should be the case: why does the virus work in most cell types but does not work in these neuronal cells?

In a lytic infection where the virus is growing, a series of events is observed in which one set of genes is expressed and their products then activate another set of genes and eventually cell lysis occurs right at the end (*Figure 5*). The critical stage in this process is the first stage in which the immediate–early genes are expressed immediately after the virus has got into the cell and it is this stage which is blocked in latency (Katz *et al.*, 1990; Margolis *et al.*, 1992; Valyi-Nagi *et al.*, 1991; *Figure 5*). Thus, the immediate–early RNAs derived from these genes are readily detectable in lytically infected permissive cells, but in latently infected neuronal cells they are not observed (Croen *et al.*, 1987; Speck and Simmons, 1991). We can say, therefore, that this is the block in neuronal cells: the virus gets in but it cannot transcribe its immediate–early genes and so the lytic cycle cannot proceed (*Figure 5*). This is why Acyclovir has no effect in latent infections because, although it is brilliantly effective at blocking DNA replication in the lytic cycle in permissive cells, it does not do anything at the immediate–early stage because this is much earlier in the cycle. Therefore Acyclovir is not an effective treatment for latent infection.

The first thing to try and understand is what goes on in cells where the immediate–early genes are expressed and why that does not happen in neurons. Many studies have defined the fact that there is a recognition sequence in the promoters of the immediate–early genes which is

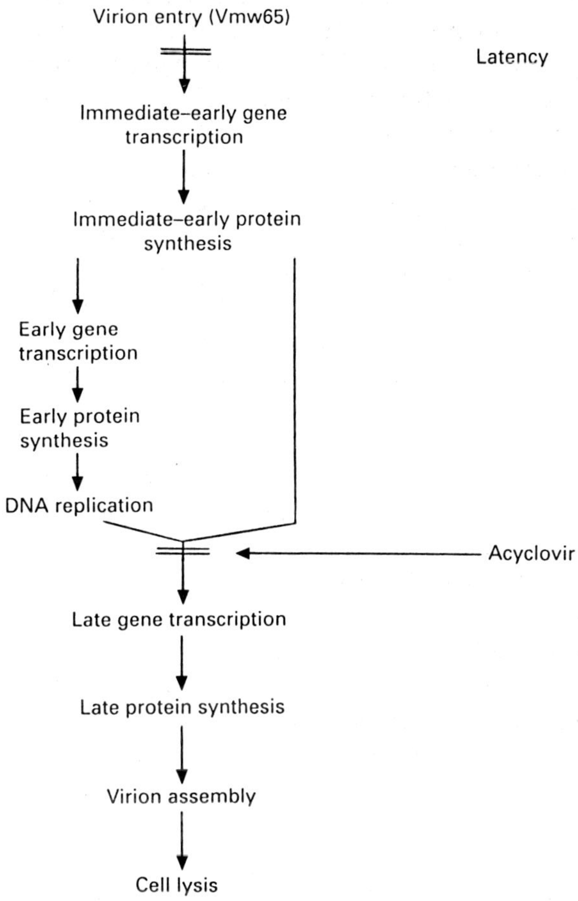

Figure 5. The lytic cycle of HSV. Note the inhibition of viral immediate–early gene transcription in neuronal cells and the blockage of viral DNA replication by the drug Acyclovir.

called the TAATGARAT sequence (*Figure 6*). This sequence is the target for a virus protein (Vmw65) which was made in the previous infected cell and enters the cell with the virus particle, and which then switches on these genes through this sequence. However, this protein does not act alone. In fact, as shown by Preston *et al.* (1988) and O'Hare and Goding (1988), it needs to make a complex with a pre-existing cellular factor in order to work, and the complex with the cellular factor can then bind to the TAATGARAT sequence and switch the gene on. Once a cellular factor enters the equation, so that cells where the lytic cycle occurs, such as the permissive cell in the skin, need to have this factor, then it can be argued that perhaps a neuron does not have it or maybe the neuron has some different factor and so on. It is difficult to see how the viral protein alone can interact differently with different cells, but once the involvement of the cellular factor is recognized, then it is possible to work out what is going on.

Obviously, the cellular factor needs to be identified, and O'Hare and Goding (1988) have identified it as an octamer-binding protein known as 'Oct-1', so called because it binds to an 8 bp sequence (*Figure 6*). In uninfected cells this factor pre-exists, switching on genes with this particular octamer sequence (*Figure 6*). However, the virus hijacks the protein and forces it to bind to this related but different TAATGARAT sequence in the virus genes. This is therefore an illustration of the parasitic nature of the virus. The next step is to investigate the process in neuronal cells: is this factor absent or is there something else there, and so on?

One difficulty is that a lot of material is needed to do these studies; they cannot be carried out on small amounts of neuronal cells dissected

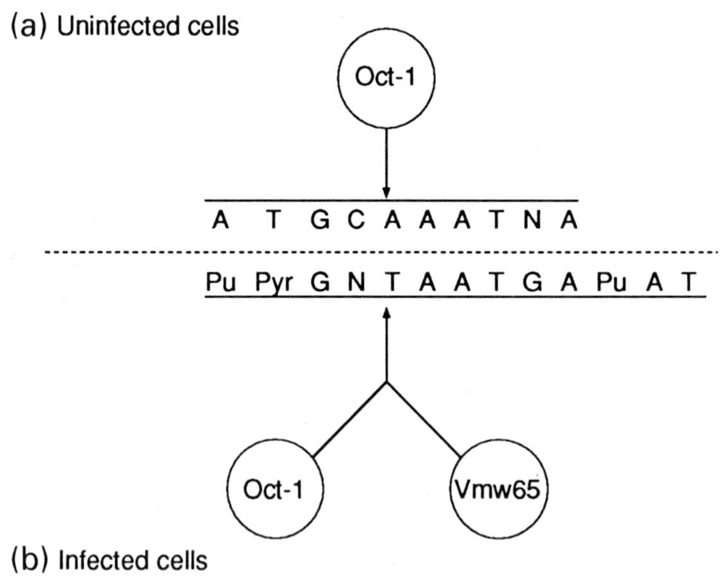

Figure 6. (a) In uninfected cells, the Oct-1 transcription factor binds to the octamer motif (ATGCAAAT). (b) In HSV-infected cells, the complex of Oct-1 and the viral factor Vmw65 binds to the related TAATGARAT motif in the viral immediate–early genes.

out of the animal, because not enough material is available. We therefore began a collaboration with John Wood who had fused primary sensory neurons from dorsal root ganglia, which is the site where the virus goes latent, with a dividing tumour neuroblastoma cell. After this fusion, a drug was added which would kill the neuroblastoma on its own so it cannot grow in the presence of the drug. Similarly, the non-dividing primary neurons cannot grow and divide. Therefore, only a hybrid which has taken some contribution from the drug-resistant non-dividing cell and put it into the neuroblastoma can grow (Wood *et al.*, 1990). In this way, a model system can be created to look at and then, as described later, one may ultimately go back at the end and try to see whether what has been learned in this system also applies to the real cells; using the smaller amounts of material available.

When looking at these neuronal cell lines, the first question asked was: do these cells support the lytic cycle or are they blocked, like real neurons? It was shown that, following infection of these neuron-derived cells, transcription of the five immediate–early genes of the virus is not seen, whereas it is in permissive fibroblasts (Wheatley *et al.*, 1990). These cells are therefore blocked in immediate–early gene transcription exactly as are real latently infected cells *in vivo*, so these cells can be used as a model system to start to learn about what is going on with the virus.

The first thing that one would like to know then is: what proteins bind to the octamer or TAATGARAT sequence in these cells? To establish this, a DNA mobility shift assay was performed in which a piece of DNA was mixed with total cellular protein. If the protein binds to the DNA, then the DNA complex is larger than the DNA alone. Therefore, the DNA moves more slowly when it is run on a gel. This is called a 'DNA mobility shift assay' because the DNA is shifted up the gel if the protein binds to it. If this is done with the octamer or TAATGARAT sequences in permissive cells such as 3T3 fibroblasts or epithelial cells which can support the virus, a single band is seen which is Oct-1, the protein described earlier (*Figure 7*). In contrast, in ND7 neuronal cells two bands are seen. Thus, in addition to Oct-1, a major complex is shown which is smaller and much more abundant and this is formed by a protein called Oct-2 (*Figure 7*). Therefore, there are two proteins present in neuronal cells, it is not that Oct-1 is not here, but that there seems to be an extra protein present (Kemp *et al.*, 1990; Wheatley *et al.*, 1991).

On the basis of this finding and a great deal of functional data, a model was proposed in which Oct-1 in permissive cells forms a complex

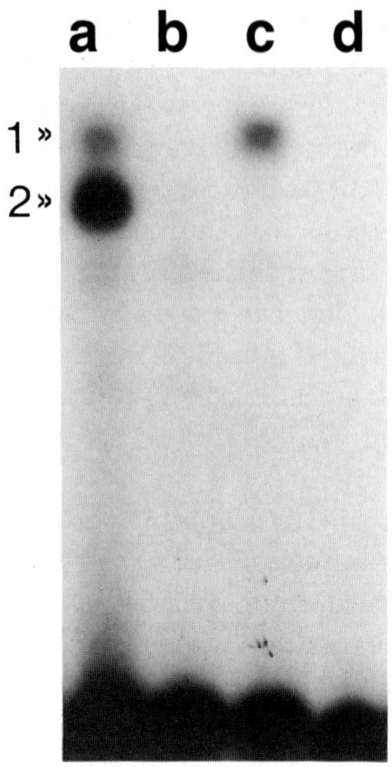

Figure 7. DNA mobility shift assay using whole-cell extracts from ND7 and 3T3 cells. Lanes a and b, ND7 cells; lanes c and d, 3T3 cells. In lanes a and c the probe was the octamer/TAATGARAT sequence ATGCTAATGATAT; in lanes b and d the probe was the mutant octamer ATAATAATAA. The arrows indicate the positions of Oct-1 and Oct-2.

with Vmw65; the complex binds to the TAATGARAT sequence and switches gene expression on. It is known that this complex forms just as well in neuronal cells: Oct-1 is already present and Vmw65 comes in with the virus, then the complex is formed in exactly the same way. However, this extra protein, Oct-2, is also present. The work of other people has shown that this Oct-2 protein does not interact with Vmw65 and does not form a complex because of differences between Oct-1 and Oct-2 (Gerster and Roeder, 1988). Therefore, the protein sits on the TAATGARAT sequence and blocks the binding of the activator complex of Oct-1 and Vmw65. It is not that the activator complex is not there, but that the Oct-2 factor actually gets there, gets in the way and blocks the binding. A great deal of evidence at this time pointed to this as being the way in which the failure of immediate–early gene expression in neuronal cells worked (Dent *et al.*, 1991; Lillycrop *et al.*, 1991).

There was only one trouble with this theory. Although it is entirely consistent with the data obtained, unfortunately it seemed to be completely inconsistent with everything that is known about Oct-2 from the work of others. Thus, although it is a protein which is absent from most cells, such as epithelial cells in the skin, fibroblasts etc., it is present in B lymphocytes circulating in the blood. Moreover, it has been shown that Oct-2 in B lymphocytes is involved in activating the immunoglobulin genes for the antibody molecules which contain an octamer sequence in their promoters. So we have a paradox: Oct-2 is a strong activator molecule switching on gene expression in B lymphocytes, yet it was proposed as an inhibitor, a molecule which gets in the way of Oct-1 and represses the immediate–early genes in neuronal cells. The question, of course, is: why does it not switch on the immediate–early genes? It could be argued that Oct-2 blocks the binding of Oct-1/Vmw65, but why does this switch off immediate–early gene expression – if it can switch on other genes, why does it not also switch on these genes? At this point, it was assumed that there must be a difference between Oct-2 in B cells and in neuronal cells but the nature of this difference was unknown.

Work by Thomas Wirth in Zurich (Wirth *et al.*, 1991) resolved the problem. It was shown that there are six different forms of Oct-2 and these forms are created by a process termed 'alternative splicing'. In B cells, on which Wirth was working, Oct-2.1 was the major form which had previously been defined there and the other five forms were minor forms in B cells. This was interesting because it was reasoned that if these other forms exist, then maybe they are more abundant in neuronal cells and therefore play some role in the difference uncovered in the activity of Oct-2 in B cells and neuronal cells.

For this reason, the levels of each of these different forms were measured in a variety of neuronal and B cell lines. It was found that the two forms, Oct-2.4 and -2.5, which are very low in B cells, are major forms in neuronal cells and Oct-2.1 is a minor form in neuronal cells where Oct-2.4 and -2.5 predominate. The two forms of Oct-2.4 and -2.5 have changes at the carboxy-terminus compared to Oct-2.1. This is important as the activation domain which results in Oct-2.1 being able to activate the immunoglobulin genes is in this carboxy-terminal region, so it was reasoned that if this region is altered then this could actually result in the conversion from activator to apparent inhibitor. In contrast, the 2.2 and 2.3 forms are not affected in the carboxy-terminus and so they should be the same as 2.1.

The next step was to take each of these forms and introduce it into permissive fibroblast cells with the virus immediate–early promoter driving a marker gene. It was seen that the three forms, Oct-2.1, -2.2 and -2.3, which have the intact carboxy-terminus, can stimulate the immediate–early promoter in the permissive fibroblast cell, whereas Oct-2.4 and -2.5 can repress it. Thus, the ability of Oct-2 to repress the immediate–early promoter in neuronal cells is controlled by the presence of these two specific inhibitory forms, whereas the forms in B cells do activate genes. The paradox is therefore resolved: it is simply that Oct-2 is different in the forms that predominate in neuronal cells compared to the B cells (Lillycrop and Latchman, 1992).

If an examination is made of the various octamer-binding proteins which exist (*Figure 8*) it is seen that Oct-1 is quite weak in its ability to switch genes on but this is unimportant because it can interact with the virus protein Vmw65 and form a complex which strongly switches on gene expression. B cell Oct-2, in contrast, cannot interact with Vmw65. However, because it can activate gene expression on its own it does not need Vmw65. In contrast, neuronal Oct-2, Oct-2.4 and Oct-2.5 have the worst of both worlds. Like all forms of Oct-2, they cannot interact with Vmw65 and so bring the strong activator to the DNA, but they are also unable to activate on their own because they are different in the carboxy-terminus activation domain.

We can now, therefore, refine our original model (*Figure 9*). As before, the interaction of Oct-1/Vmw65 switches on genes in permissive cells and that interaction still happens in neuronal cells. However, neuronal Oct-2 is incapable of interacting with Vmw65 and is also unable to activate on its own. It therefore binds to TAATGARAT and inhibits gene activity by preventing binding of the Oct-1/Vmw65. We now have a form of Oct-2 which does exactly as predicted and represses the viral immediate–early genes. Of course, there is one more prediction which can be made: if this is true and the presence of this neuronal Oct-2 is enough to explain the failure of the virus to grow in our neuronal cells, then it should be possible to express Oct-2 artificially in a cell where the virus will normally work and stop the virus growing in those cells. Thus, if B cell Oct-2 is introduced into a cell which only has

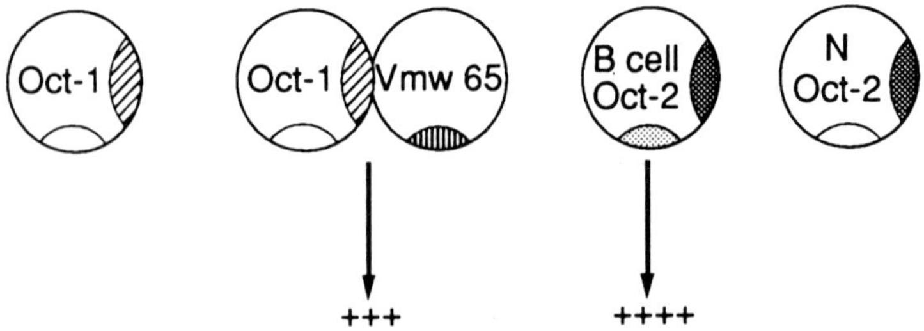

Figure 8. Octamer-binding proteins. Note the different regions which allow Oct-1 and not Oct-2 to interact with Vmw65 and the strong activating regions of B cell Oct-2 and Vmw65 which are absent in Oct-1 and neuronal Oct-2.

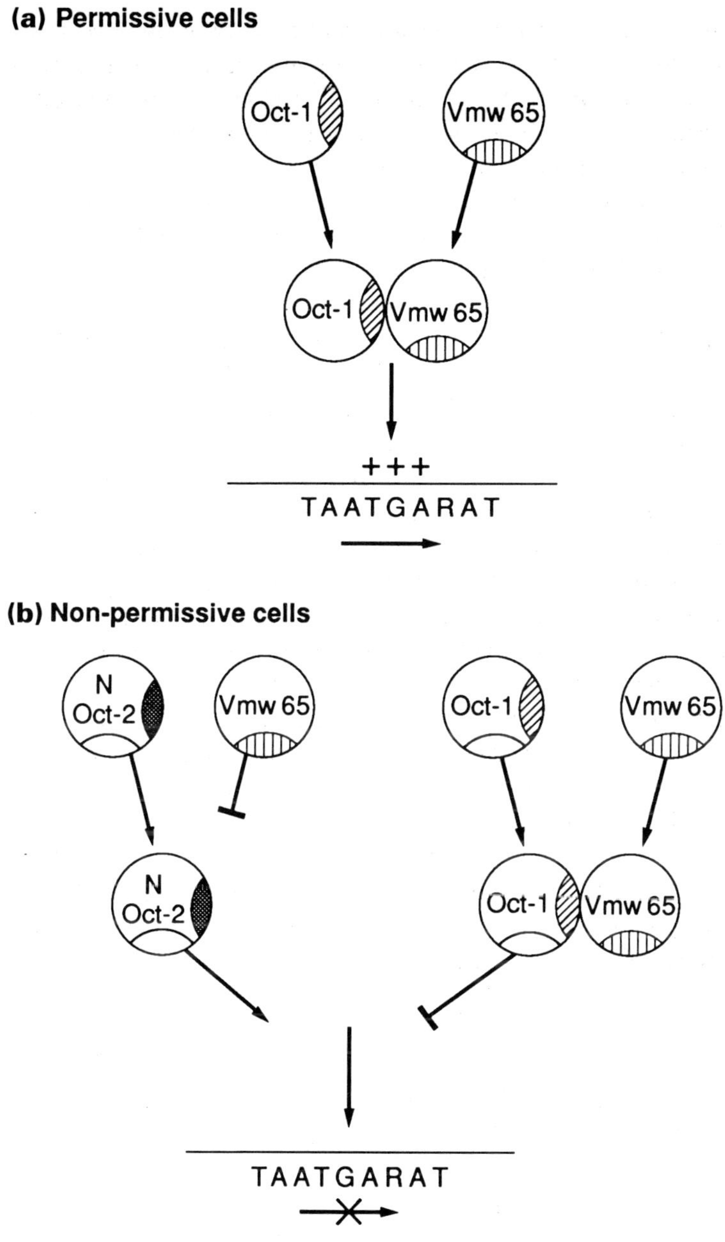

Figure 9. (a) In permissive cells, the complex of Oct-1 and the strong activator Vmw65 activates the HSV immediate–early genes by binding to the TAATGARAT sequence. (b) In non-permissive neuronal cells, this complex still forms but is prevented from binding by neuronal Oct-2 which therefore prevents the activation of the immediate–early genes.

Oct-1 and in which the virus can grow, the virus should still be able to grow in those cells. In contrast, if neuronal Oct-2 is expressed, it should be possible to make those cells non-permissive for the virus.

In fact that is exactly what happens. Normally permissive baby hamster kidney (BHK) cells expressing the neuronal Oct-2 or B cell Oct-2 were produced and then infected with HSV. An assay for the transcription of the five immediate–early genes showed a dramatic fall in the transcription of the five immediate–early genes in the cells expressing Oct-2.4 or -2.5 but no effect in the cells expressing Oct-2.1. Thus, expressing these neuronal transcription factors in normally permissive cells blocks immediate–early gene transcription.

As shown earlier, immediate–early gene transcription is critical for the rest of the virus cycle. Of course, now we actually look at how much virus these cells produce when they are infected with HSV. The production of virus on addition of a small amount of virus to each of these cells was measured and then the virus that came out was harvested, so testing whether the virus can replicate in the cell. In fact, in these experiments, in the cells which express the B cell form of Oct-2, a slight enhancement of viral production is found, but in the cells with the neuronal forms of Oct-2.4 or -2.5 the permissiveness of these cells and their ability to produce virus can be dramatically reduced. The behaviour of a neuronal cell in response to HSV can therefore be mimicked simply by transferring a single neuronal transcription factor over into a cell in which the virus normally grows. In this experiment, the reason why our neuronal cells are incapable of replicating the virus has been demonstrated. It is simply because they express high levels of Oct-2.4 and -2.5.

The next step was to establish: if Oct-2 explains this phenomenon for these cell lines, does it explain it for real neurons as well? Is Oct-2 also actually present in real neurons? By carrying out *in situ* hybridization experiments, the presence of Oct-2 in real sensory neurons was demonstrated. The Oct-2 RNA was observed both in neurons dissected directly from the animal and also those maintained in culture. So Oct-2, which we originally discovered in the neuronal cell line, is also present in the real site of herpes latency – the site where the virus does not replicate and establishes its latent infection (Lillycrop *et al.*, 1991).

Some interesting experiments could then be performed with real neuronal cells to discover whether factors that affect latency in these cells also affect Oct-2. If factors are added which regulate latency, do they regulate our factor which is believed to be responsible for latency in these cells? To test this, nerve growth factor (NGF) was added to these cells, since NGF is important for the maintenance of latency – without NGF, latency cannot be maintained (Wilcox and Johnson, 1988). It was shown in these experiments that both the Oct-2 RNA and the Oct-2 protein increase in abundance following treatment with NGF (Wood *et al.*, 1992).

Thus, Oct-2 is actually stimulated by treatment with NGF. This returns to the experiment of Cushing in 1905, because NGF is made by the peripheral tissues and transported back to the nerve along the nerve process. When the nerve process is cut, this retrograde transport of NGF is blocked. So, the experiment which Cushing carried out 88 years ago has been explained here effectively in molecular terms; if NGF transport is blocked, Oct-2 levels will fall, the virus will reactivate and produce a lytic infection (*Figure 10*). In this way, this factor can be used to explain the phenomenon of latency.

As described initially, not just the virus but also the cell is of interest to our studies. A cellular factor has been found here which is present in neuronal cells and is not there solely to restrict the virus, since it is most unlikely that the cell evolves a factor solely for the purpose of dealing

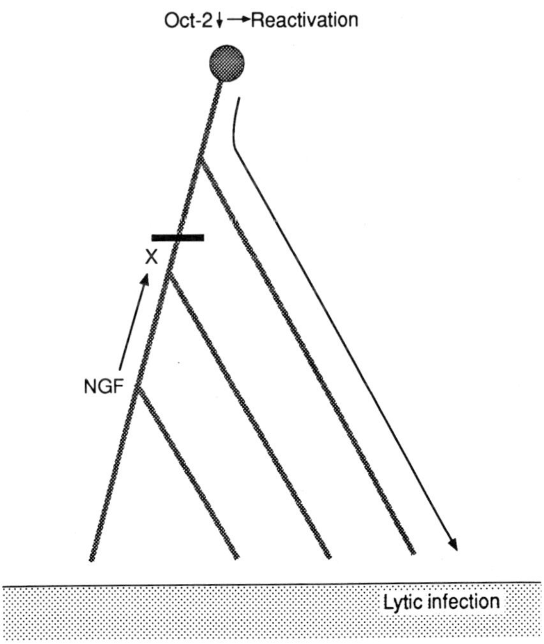

Figure 10. Cutting of the nerve process blocks the retrograde transport of nerve growth factor (NGF) and thereby leads to a fall in Oct-2 levels, resulting in viral reactivation.

with the virus. This factor is there to regulate specific cellular genes and as part of this process it also regulates the virus. Cellular genes in the neuron which Oct-2 might regulate are now being sought. A few genes with octamer-binding sites in their promoters have been identified, such as those encoding creatine kinase, calcitonin gene-related peptide (CGRP) and tyrosine hydroxylase, and their modulation by Oct-2 is currently being investigated. Interestingly, several of these genes, such as tyrosine hydroxylase and CGRP, are actually regulated by NGF. Thus, these genes are modulated in their expression by NGF – the factor which has been shown to modulate Oct-2 expression. Therefore, this may indicate a pathway whereby NGF modulates Oct-2 and that modulates the virus, but that also modulates cellular genes. This is currently being investigated.

3. Conclusions

To return to the virus, we now know that Oct-2 is important in that it prevents the lytic cycle in neuronal cells and that is the prerequisite for the establishment of latency. Does this actually help a patient who is visiting the doctor for Acyclovir to treat recurrent cold sores? Is it indeed true that: 'once bitten, forever smitten'? How could Oct-2 be used to get at the virus? Not at the onset of latency because, as shown, people become infected very early on. Similarly, it is unlikely that it is worth trying to lower Oct-2 levels and so bring the virus out of the neuron once and for all, because damage will be caused to the neuron in the process. On the other hand, it might be possible to try to keep Oct-2 levels up in order to keep the virus in a state where it cannot reactivate. This does not seem to be a viable treatment to give to people continuously. However, at critical periods where people are immunosuppressed, for example, it may be worth having a treatment which could actually keep the virus silent. Finally, experiments on Oct-2 expression in the permissive fibroblast cells represent the first in which a single cellular factor has been put into a cell and made it non-permissive, so they do open up the possibility of expressing inhibitory factors in cells that normally lack them and thereby block infection. This could be of some use in HSV, but may, of course, be of more use in other viruses, for example, HIV, which is probably regulated in part by negative factors and for which no effective treatment is available – even for lytic infections.

Hopefully this overview has shown that herpesvirus is medically interesting in itself, and is also interesting as a tool to look at the cell.

Acknowledgements

Thanks to Action Research, who have generously supported this work at a time when it was a bit of a 'shot-in-the-dark' to look at viruses in this way. I am also very grateful to Susan Wheatley, Lynn Kemp, Carolyn Dent, Karen Lillycrop, John Estridge and Keith Howard in my laboratory, who have worked with herpes, and also those who have worked on all the other projects, many of them collaborative, such as heat-shock proteins and RNA splicing.

References

Croen KD, Ostrove JM, Dragovic LJ, Smialek JE and Straus SE. (1987) Latent herpes simplex virus in human trigeminal ganglia: detection of an immediate–early gene anti-sense transcript by *in situ* hybridization. *N. Engl. J. Med.* **317**, 1427–1432.

Cushing H. (1905) Surgical aspects of major neuralgia of trigeminal nerve: report of 20 cases of operation upon the Gasserian ganglion with anatomic and physiologic notes on the consequences of its removal. *J. Am. Med. Assoc.* **44**, 1002–1008.

Dent CL, Lillycrop KA, Estridge JK, Thomas NSB and Latchman DS. (1991) The B cell and neuronal forms of the octamer binding protein Oct-2 differ in DNA binding specificity and functional activity. *Mol. Cell. Biol.* **11**, 3925–3930.

Gerster T and Roeder RG. (1988) A herpes virus transactivating protein interacts with transcription factor OTF-1 and other cellular proteins. *Proc. Natl Acad. Sci. USA*, **85**, 6347–6351.

Katz J, Bodin T and Coen DM. (1990) Quantitative polymerase chain reaction analysis of herpes simplex virus DNA in ganglia of mice infected with replication incompetent mutants. *J. Virol.* **64**, 4288–4295.

Kemp LM, Dent CL and Latchman DS. (1990) Transcriptional repression of herpes simplex virus immediate–early genes and of octamer-containing cellular promoters in neuronal cells is mediated by the octamer DNA element. *Neuron*, **4**, 215–222.

Latchman DS. (1990) Molecular biology of herpes simplex virus latency. *J. Exp. Pathol.* **71**, 133–141.

Lillycrop KA, Dent CL, Wheatley SC, Beech MN, Ninkina NN, Wood JN and Latchman DS. (1991) The octamer binding protein Oct-2 represses HSV immediate–early genes in cell lines derived from latently infectable sensory neurons. *Neuron*, **7**, 381–390.

Lillycrop KA and Latchman DS. (1992) Alternative splicing of the Oct-2 transcription factor RNA is differentially regulated in neuronal cells and B cells and results in protein isoforms with opposite effects on the activity of octamer/TAATGARAT-containing promoters. *J. Biol. Chem.* **267**, 24960–24966.

Margolis TP, Sedarati F, Dobson AT, Feldman LT and Stevens JG. (1992)

Pathways of viral gene expression during acute neuronal infection with HSV-1. *Virology*, **189**, 150–160.

O'Hare P and Goding CR. (1988) Herpes simplex virus regulatory elements and the immunoglobulin octamer domain bind a common factor and are both targets for virion transactivation. *Cell*, **52**, 435–445.

Preston CM, Frame MC and Campbell MEM. (1988) A complex formed between cell components and a herpes simplex virus structural polypeptide binds to a viral immediate–early gene regulatory DNA sequence. *Cell*, **52**, 425–434.

Roizman B and Sears AE. (1987) An inquiry into the mechanisms of herpes simplex virus latency. *Annu. Rev. Microbiol.* **41**, 543–571.

Speck PG and Simmons A. (1991) Divergent molecular pathways of productive and latent infection with a virulent strain of herpes simplex virus type 1. *J. Virol.* **65**, 4001–4005.

Stevens JG. (1975) Latent herpes simplex virus and the nervous system. *Curr. Top. Microbiol. Immunol.* **70**, 31–50.

Valyi-Nagi T, Deshmane SL, Spivack JG, Steiner I, Ace CI, Preston CM and Fraser NW. (1991) Investigation of herpes simplex virus type 1 (HSV-1) gene expression and DNA synthesis during the establishment of latent infection by an HSV-1 mutant in1814 that does not replicate in mouse trigeminal ganglia. *J. Gen. Virol.* **72**, 641–649.

Wheatley SC, Kemp LM, Wood JN and Latchman DS. (1990) Cell lines derived from dorsal root ganglion neurons are non-permissive for HSV and express only the latency-associated transcript following infection. *Exp. Cell. Res.* **190**, 243–246.

Wheatley SC, Dent CL, Wood JN and Latchman DS. (1991) A cellular factor binding to the TAATGARAT DNA sequence prevents the expression of the HSV immediate–early genes following infection of non-permissive cell lines derived from dorsal root ganglion neurons. *Exp. Cell. Res.* **194**, 78–82.

Wilcox CL and Johnson EM. (1988) Characterization of nerve growth factor-dependent herpes simplex virus latency in neurons *in vitro*. *J. Virol.* **62**, 393–399.

Wirth T, Priess A, Annweiler A, Zwilling S and Oeler B. (1991) Multiple Oct-2 isoforms are generated by alternative splicing. *Nucleic Acids Res.* **19**, 43–51.

Wood JN, Bevan SJ, Coote P *et al.* (1990) Novel cell lines display the properties of nociceptive sensory neurons. *Proc. R. Soc. Ser. B*, **241**, 187–194.

Wood JN, Lillycrop KA, Dent CL, Ninkina NN, Beech MN, Willoughby JJ, Winter J and Latchman DS. (1992) Regulation of expression of the neuronal POU protein Oct-2 by nerve growth factor. *J. Biol. Chem.* **267**, 17787–17791.

Index

ALSO AVAILABLE FROM BIOS SCIENTIFIC PUBLISHERS LTD

Autoimmune Disease

Focus on Sjögren's Syndrome

D.A. Isenberg & A.C. Horsfall (Eds)
respectively University College, London, UK; and Kennedy Institute of Rheumatology, London, UK

This book is a comprehensive guide to many of the factors that are thought to be involved in the aetiology of autoimmune diseases in general, and in Sjögren's Syndrome in particular. Sj ögren's Syndrome is believed to be the commonest autoimmune rheumatic disease and a paradigm for most autoimmune disorders. A key theme of the book is the application of molecular biology to increase our knowledge of the structure and function of genes. The products they encode play integral roles in the development of these, so far, incurable diseases.

Contents

Of interest to:

Clinicians and researchers.

Series: UCL Molecular Pathology

Hardback 240 pages 1-872748-23-6 July 1994

The Chromosome

ORDERING DETAILS

Main address for orders

BIOS Scientific Publishers Ltd
St Thomas House, Becket Street,
Oxford OX1 1SJ, UK
Tel: +44 1865 726286
Fax: +44 1865 246823

Australia and New Zealand
DA Information Services
648 Whitehorse Road, Mitcham, Victoria 3132, Australia
Tel: (03) 873 4411
Fax: (03) 873 5679

India
Viva Books Private Ltd
4346/4C Ansari Road, New Delhi 110 002, India
Tel: 11 3283121
Fax: 11 3267224

Singapore and South East Asia
(Brunei, Hong Kong, Indonesia, Korea, Malaysia, the Philippines, Singapore, Taiwan, and Thailand)
Toppan Company (S) PTE Ltd
38 Liu Fang Road, Jurong, Singapore 2262
Tel: (265) 6666
Fax: (261) 7875

USA and Canada
Books International Inc
PO Box 605, Herndon, VA 22070, USA
Tel: (703) 435 7064
Fax: (703) 689 0660

Payment can be made by cheque or credit card (Visa/Mastercard, quoting number and expiry date). Alternatively, a *pro forma* invoice can be sent.

Prepaid orders must include £2.50/US$5.00 to cover postage and packing for one item and £1.25/US$2.50 for each additional item.